WHAT IF YOU'RE WRONG?

The Dire Cost of Mistaken Beliefs

by Adrian Milton

Copyright © 2013

CONTENTS

PREFACE

I'll admit it: **I believe some really bizarre things.** I won't tell you what they are just yet, but I can fully understand why some people would find my beliefs hard to swallow. We ***all*** have some pretty strange beliefs. After all, the world is a pretty strange place. From the singularity of a black hole to the duality of light, from the first appearance of life on Earth to the billions of wondrous creatures that have come and gone over the eons, the world is nothing if not unbelievable.

I used to maintain the philosophy, "Don't question my beliefs and I won't question yours," but I've come to realize that it's actually *healthy* to question one's own (and other people's) beliefs. That hit home this past spring when I noticed a strange splotch on my mother's cheek. She believed it was nothing. An old person's skin, she said, is chock full of odd moles. But I questioned her belief. I showed her what I found on SkinCancer.org and convinced her that a visit to the dermatologist was warranted. And I was right. She had three patches of basal cell carcinoma which were promptly removed. Healthy indeed to question a person's beliefs.

As important as my mother's well-being is to me, this book is more about questioning what we might call Big Beliefs...and I'm not just referring to religious ones. From the Big Bang to Bigfoot, Area 51 to 9/11 conspiracies, there are a whole lot of strange ideas out there that a whole lot of people seem to believe in. (Again, I include myself among their number.) And so, maintaining the stance that *it is beneficial to question strange beliefs,* I present to you this book. If it creates even the tiniest crack in the armor of certainty most people have about their beliefs, it will have been worth it.

CHAPTER 1:
It's Okay to Question Your Beliefs

INTRODUCTION:

I'd like to start by giving you my warmest welcome. Think of the open cover of this book as an open *door*, instead. I'm holding it open, gesturing for you to come in and have a seat. I'd like you to be comfortable because we're going to have a discussion, you and I. A discussion about our beliefs.

The purpose of this book is to help you explore your beliefs and maybe, just maybe, get you to reexamine a few of them. I'm not hoping for, well, a *miracle*. I don't expect anyone with strong beliefs to suddenly change them after reading one book. And that, of course, begs the question: **Then why bother trying?** Why should I waste my breath trying to change someone's belief when I feel certain that I won't succeed?

The answer is: There's so much at stake for both of us.

At this point you might be wondering: *Wait....Who is this guy? Is he an Evangelist trying to reach out to an atheist reader? Or just the opposite: An atheist who actually thinks he can get me to turn my back on God? (Good luck, buddy!)*

Okay, I'll fess up: I categorize myself as being rational. I try to judge all statements by applying logic to determine truth, or the likelihood of it. In short, I aspire to use reason as my source of knowledge.

But enough about me. You're my guest, and we're both here to discuss our beliefs. So here's how this works: This book is arranged as a series of statements. After reading the first statement, you decide whether you AGREE or DISAGREE. Then, depending on your choice, you will be directed to read one particular section or another, for a discussion about your belief (or disbelief, as the case may be).

For example, a statement might be: **Science is the best tool we have to help us understand our world.**

You then decide: **AGREE** and then read the section that follows.

...or...

DISAGREE and read that section instead.

Be aware, sometimes the "AGREE" discussion will come first, and other times the "DISAGREE" section will come first. If you read both points of view, as I hope you will, then you'll see that the particular arrangement of the two sections is done to facilitate the flow of the overall discussion about the statement itself.

Also, some statements that seem like obvious facts to you will nevertheless seem like bizarre beliefs to someone else.
For example:

The Bible is the unerring word of God.

The universe was once smaller than a grain of sand.

The purported moon landing in 1969 was actually a hoax perpetrated by the U.S. government.

Other statements, on the other hand, probably won't be so clear cut for you:

Cellphone usage can cause brain tumors.

Eyewitness accounts are sufficient evidence to declare something as fact.

The presence of highly specified complexity in any object or system always indicates an intelligent designer.

So, although it's fine to read a statement and then jump to the particular section that matches your belief (i.e. Go directly to the "AGREE" section or to the "DISAGREE" section), I encourage you to read *both*. After all, you're always better off knowing both sides of an issue.

One thing to note: This isn't a book that debunks beliefs. For example, if you agree with the statement that **"The Great Pyramids were built by aliens,"** I'm not going to quote the voluminous scientific evidence in an effort to convince you otherwise. Instead, through questions and examples, I'll try to get you to reexamine your belief in the hope that you might reconsider it. This book is more about getting you to think for yourself than it is about battering you with facts.

I mentioned above that I'm writing this book because there's so much at stake for both of us, and that is absolutely true. Every single action you take in life—from the most mundane to the most vital—is directly affected by your beliefs. You might believe a certain new diet is healthy for you (and so you take the action of adhering to the diet), when in reality it's harmful.

You might believe adults can't learn a second language and so you turn down an offer for a new job that requires foreign language training. You might believe your betting strategy can beat the odds at the roulette table, and you end up losing a significant amount of money. You might believe it's impossible to lose a significant amount of weight, and so you never try.

When we're talking about beliefs, it's fair to say that *your whole life is at stake*. So if you're ready to reexamine some of your own beliefs, then please go on to the first statement.

STATEMENT: The earth is round.

AGREE:
Why do you believe the earth is round?

It may seem like a frivolous question, but it's actually an important one, and worth asking yourself honestly. And it's not good enough to say, "Everyone knows it's round." After all, a few thousand years ago, everyone "knew" the earth was flat. So I ask again: Why do you believe the earth is round? Did some photos in science class convince you? Or is it just that everyone around you refers to it as being round?

One way to answer is to imagine being transported back to a time when everyone believed the earth to be flat. Pretend you don't have any satellite photos to show them. Just words. How would you go about convincing these people that the earth is spherical? (It's not so easy, is it?) Obviously I'm not trying to get you to reconsider your belief that the earth is round. I just wanted to start our discussion with a bit of critical thinking as a kind of warm up for what's to come. And I wanted to emphasize the idea that *every now and then it's a good idea to question even your most fundamental beliefs.*

Before leaving this topic, I have an open-ended question for my religious readers to consider: If your holy book—for example the Bible or the Koran—stated that the earth was flat, would you change your belief? (People usually get defensive on that one. "It doesn't say that!" they insist. But that's sidestepping the question.)

DISAGREE:
Wow! You're an actual member of the Flat Earth Society? (Apparently, as of 2012, their group is 420 members strong, at least according to Wikipedia.) I actually respect the consistency of Flat Earthers. In their interpretation the Bible tells them that the earth is flat, so the earth is flat. Period. They are the ultimate Christians in that regard. I've spoken with some of the most fervent Southern Baptists, but even *they* don't go that far. Perhaps they accept it grudgingly, but all the Christian fundamentalists I've ever met told me they accept that the earth is round.

And yet here I am, talking with a Flat Earther. I have many questions for you, like: Do you have a photo of the edge of the earth? Can you take me there? What lies beyond the edge? and so on. But I'll limit myself to just this: Why do you think everyone else considers the earth to be shaped like a sphere?

Finally: It's okay to change your mind and accept that the earth is round. Consider speaking to your local Baptist preacher about your belief. As I mentioned above, even *he* will explain that the earth is round.

STATEMENT: Everything we believe is merely an opinion or a feeling, otherwise we wouldn't believe it, but instead we'd *know* it.

AGREE:

I'm having trouble accepting that *everything* I believe is merely an opinion. When I state that, "I believe the earth is round," is that really just my opinion? You might argue that, yes, ultimately it's still my opinion, albeit one that's shared by virtually every human being on the planet, and backed by overwhelming physical evidence. You might also argue that, since it was a statement of fact, I'm incorrectly using the word "believe" when I should be using the word "know."

What's really at issue here is that the word "believe" has two very different meanings:

1) **Believe = feel; "my opinion is"**

Examples:
I **believe** the Yankees should improve their bullpen.
(**In my opinion**, the Yankees should improve their bullpen.)

8

I believe we should go to Cancun for our vacation.
(**In my opinion**, we should go to Cancun for our vacation.)

2) **Believe = know; "accept as fact"**

<u>Examples</u>:
I **believe** the earth orbits the sun.
(I **accept it as fact** that the earth orbits the sun.)

I **believe** smoking causes lung cancer.
(I **accept it as fact** that smoking causes lung cancer.)

Those are two completely contradictory meanings. (*)
"Believe" can indicate someone's subjective opinion, or it can indicate an objective fact, and yet people use these two meanings interchangeably. Or at the very least, people often use the word without being clear which meaning they have in mind.

How about you? Which meaning would you use for the following?

- I **believe** that Dunkin Donuts makes better coffee than Starbucks. (FEEL? or KNOW?)

- I **believe** that light travels faster than sound. (FEEL? or KNOW?)

Now, if you're religious, which meaning would you use for this next one?

- I **believe** God exists. (FEEL? or KNOW?)

To me, that's the difference between religious moderates versus religious fundamentalists. When moderately religious people believe in God, they *feel* it in their heart (whatever that actually means). They don't force their opinions on other people, just as

9

I don't go around trying to convince everyone to drink Dunkin Donuts coffee. But when a fundamentalist believes in God, he **knows** it to be a fact. And thus he sets out to aggressively preach "the truth." These two groups are using different meanings of the same word. I'm totally supportive if you want to "feel" that your God exists. But I (and all rational thinkers) have a big problem when you purport to **know** a god exists.

The bottom line is: We need to realize that **much of what we believe is merely an opinion or a feeling, and worthy of close scrutiny**.

DISAGREE:

This is a tough one and comes down to semantics. As we saw in the 'AGREE' discussion above, the confusion is due to the two contradictory meanings of "believe," namely: **to feel** versus **to know**. Complicating matters further is the word "belief." This may seem like it's covering the same issue, but it's not. The difference is that "believe" is a verb...it's an **action**. I look at the looming thunderclouds overhead and actively decide, "I believe it's going to rain." A belief, on the other hand, is a **mental state**. I look at the looming thunderclouds overhead and tacitly assume they're **real** and not, say, a holographic projection on some massive screen. My brains holds the belief that the clouds are real, but *it's not anything I actively reasoned*. As it turns out, the majority of our **beliefs** simply exist, without having been actively reasoned.

Due to how we usually use the word "belief," it almost always refers to something we accept to be true. *A belief, then, is a conscious or unconscious acknowledgment that something exists or is true without needing further proof.*

Notice the word "unconscious" in the above definition. The bulk of our beliefs, like that of the thunderclouds being real, are unspoken. When you drive into your garage, your brain is unconsciously adhering to the belief that your house won't suddenly come crashing down on you. (Otherwise, you obviously wouldn't drive inside.) But I doubt you consciously weigh that belief each time you come home. When you turn on your computer, it's your brain's belief that it's actually *your* computer, and not a look-alike computer that the FBI secretly put in its place when you were gone. When you drive to the mall, you're expressing your tacit belief that *it's in the same location that it was last week* when you were there, and so on. Our brain has to assume all sorts of beliefs like these—countless thousands of them—or we'd never be able to function. And none of them are considered "opinions" but instead facts about the world.

Still, it's a person's *conscious* beliefs I'm more concerned about; the ones we make when we weigh the available evidence and decide, "Yes, this is my belief." Although our unconscious beliefs can be fun to examine (and they explain lots of interesting things, like why we fall for illusions), this book concentrates all its energy on those relatively few yet critically important *conscious beliefs* we come to hold.
If you're open to taking a closer look at some of them, then read on.

(*) **Footnote**
In the spirit of thoroughness, let's note that the phrasal form of the verb, "to believe **in**" adds yet more meanings (and confusion).

"believe in" = "value"
Example: I **believe in** hard work. (I **value** hard work.)

"believe in" = "to be convinced that someone/something can succeed"
Example: I **believe in** our pitching staff this season. (I'm **convinced** our pitching staff **can succeed** this season.)

"believe in" = "to be convinced that something exists"
Example: I **believe in** love at first sight. (I'm **convinced** love at first sight **exists**.)

STATEMENT: Human beings are not actually alive, but instead an emergent system composed of trillions of living cells.

DISAGREE:
"That's ridiculous!" is what I'm assuming most readers will say. "Of course we're alive. It's virtually self-evident. I consume and burn energy, I grow, I respond to stimuli, I have the ability to reproduce, etc."

I take issue with each of those claims, but I'll start with that last one. True, humans have the ability to reproduce, but reproduction is very different from *self-replication*. You do not, in fact, create exact copies of yourself when you have children, do you?

"So?" responds the indignant reader. "Living things reproduce either asexually (by creating an exact copy of themselves), or sexually (by mixing their genes with another member of their species.)"

I'm sorry, but now you have me confused about the word "reproduce." When we talk about a reproduction—say, of a painting—***aren't we implying it's more or less identical to the original?*** Sexual "reproduction" is like cutting up both the Mona Lisa and a Van Gogh self-portrait into thousands of little pieces, then combining half the pieces from one painting and half from the other into a single, new "reproduction." It's a completely new painting that only vaguely resembles the two parent paintings. In my opinion, the process of human reproduction is a misnomer. It should be called "recombining."

I'm not saying humans aren't part of an ongoing, living process. That we indeed are. We are even living ***systems***. Still, you can definitely argue that any particular human being itself is not alive (or at the very least, not the "unit of life"), but instead is a conglomerate of trillions of microscopic organisms called cells which ***are alive***, because they ***do*** replicate themselves. Cells are the unit of life, not bodies.

Think of it this way: If you could simultaneously separate each cell in your body—put them all in separate petri dishes—you most certainly would no longer be considered alive, yet each of your cells still would be, (give or take a few dead skin, hair and nail cells).

In fact, go back and look at that default objection: *I consume and burn energy! I grow! I respond to stimuli!*

Think again. ***You*** don't consume energy at all. Your ***cells*** do. You don't grow. Your cells multiply. You don't respond to stimuli. Your nerve cells do. Frankly, I think we give ourselves way too much credit. When you get sick, it's not ***you*** fighting off the infection! Do you take any active steps to save yourself from the ravages of an invading virus? No, but your T-cells and B-cells do. You just lie there drinking tea and reading a good book.

Now, some might try to take that same argument one step further: "By that logic then, our cells are not alive because they themselves are merely conglomerates of atoms which are by no definition alive." But that's like arguing that a car can't drive because it's made of a gas pedal and pistons and spark plugs, none of which can drive. Cars *can* drive, despite their non-drivable parts, and cells *are* alive, despite their non-living parts. And cells, I maintain, are the unit of life. The emergent phenomenon of "you"-ness is simply the by-product of trillions of cells organized together, controlled by a large brain.

Think of the emergent phenomenon of an ant colony. We'd never label the colony *itself* to be "alive," would we? We realize it's the individual ants that are actually alive. In fact, they live and die, constantly being replaced while the colony continues to thrive. The colony is always changing, but its specialized ants are being replicated all the time, and each remains virtually unchanged.

Similarly, each of us is an immense colony of cells. Our cells live, work, die, and get replaced, over and over, while we continue to thrive. Your body is always changing—that is, the you of 2 years old was very different compared to the you of 12 yrs or the you of 22—but your specialized cells have remained virtually unchanged despite numerous cycles of replication.

It's easy to see that an ant colony is not actually alive because each ant is large and physically separate. It's harder to see that the colony of a human body is equally not alive, because each cell is so small, and mostly they're tightly bound together. Nevertheless, in the spirit of getting you to reexamine your most fundamental beliefs, I would have you at least contemplate that thought: You are not actually alive. Your cells are. Your sense of "you"-ness is an illusion.

AGREE:

I'm glad you showed up for this discussion. I was worried that everyone was sitting in the auditorium across the hall, taking issue with the statement that none of us are actually alive. To be honest with you, I'm not sure where I stand on that statement. I mean, I agree that each of us is indeed a colony of cells, and our feeling of self is an illusion that emerges from their complex arrangement. But, come on! Of *course* we're alive, right?

The way I see it is, it has to do with ***the scale of the observer.***

Imagine you have a device which lets you look at a planet outside of our solar system and see it very close up. As you peer into this device, the viewscreen is filled with what looks like a cell. Incredible! This alien cell appears to be using energy to maintain its stasis. As you continue to observe, you see it divide to become two new cells. This thing is definitely alive! But then you zoom out, and see it's surrounded by other cells. Soon, you've zoomed out so much, you realize you've been zoomed in on one cell of a large creature. And as you watch this creature, you see it also exhibit signs of life: When a smaller creature passes by, the larger creature follows (a reaction to stimuli). And then it engulfs the smaller creature (taking in energy to maintain its stasis), excreting the bits it can't use.

Both the cell and the creature are alive in different ways. If we all walked around with electron-microscope glasses on, zoomed in on things so much that we could see only a cell or two at a time, then we'd be more inclined to accept the statement that we really are just colonies of cells. But the reality is we're both. It's just a matter of perspective.

STATEMENT: Computer simulations show that there simply wasn't enough time for evolution to progress as quickly as it did.

DISAGREE:
I have a brilliant friend who feels that way. In his opinion, computer simulations prove that evolution could not possibly have proceeded as quickly as it did simply through random mutations and natural selection. We had a discussion about it a few months ago, the last time we hung out. I asked him, "Are these the same computer simulations that can't predict the weather three days from now? And you're asking them to calculate the path of evolution for trillions upon trillions of life-forms over the course of three billion years?"

They were powerful computers, he assures me.

I also press him on which exact computer simulation he's referring to, and who ran the studies. "There have been many simulations run," is the general reply I get.

"Okay," I tell him. "So just give me one."

He brushes the question off and I let it slide. And then I usually ask, "So, what are you postulating? In your opinion there wasn't enough time. So, how was evolution guided, exactly?"

This is where his reply gets very murky. Something about a "cosmic consciousness." He uses very big words, and sounds very convincing. You feel like you're talking to a Harvard physics professor (which is fairly close to his profession). But all it really showed me was this: ***Even brilliant people can believe some very strange things.*** As Michael Shermer explains in his insightful book, <u>Why People Believe Weird Things,</u> "Smart people believe weird things because they are skilled at defending beliefs they arrived at for non-smart reasons." (27)

My real point is this: If the most brilliant among us are not immune to forming (and maintaining) strange beliefs, then it's likely that none of us are. All the more reason we should take a little time once in a while to reexamine some of our Big Beliefs.

<u>AGREE</u>:
Will someone please let me know which computer simulations you're referring to? Can I see the study? To me, this is taking on the quality of an urban legend. "My nephew's wife has a friend whose brother works at a computer lab, and they've run the simulation dozens of times."

The belief that computer simulations somehow prove there wasn't enough time for evolution to happen the way it did is really just an ***argument from incredulity***. That is, "I'm not able to imagine how it could've happened, and so the premise must be false." And frankly, that disturbs me. You guys are supposed to be the smart ones. How can you rely on such a blatant logical fallacy?

I'm going to go out on a limb and guess that if you truly subscribe to the "computer simulations disprove evolution" argument above, then you're not an expert on the study of evolution, and do not have a degree in evolutionary biology, nor any closely related academic field. Ultimately, no matter who you are, such computer simulations (if they indeed exist) reveal nothing about the process of evolution. (Though they reveal quite a bit about the flaws of reasoning that formed the basis of the study itself.)

What's really going on here is that even brilliant people want to believe in something bigger than themselves. They realize the associations and baggage a word like "God" has, so they sidestep it with equally unfounded (but more "scientific sounding") terms like my friend's "cosmic consciousness" and Dembski's "intelligent designer." God by any other name, my friends...

STATEMENT: Our belief in gods is the result of certain unconscious mental functions.

AGREE:
I'm writing either to an evolutionary psychologist, or to someone who's well-read on the topic. Either way, I, too agree with that topic statement. The field of evolutionary psychology goes *such* a long way towards explaining the existence and persistence of religion, that it really should be required study in college. In *high-school*, even. When I was a teenager, I would've loved to use some of those terms on my parents. "Why do we have to pray before dinner all the time? Don't you guys know you're just decoupling your cognition and letting your innate theory-of-mind mechanism be fooled by your hyperactive agency detection device?"

If you read the 'DISAGREE' section below, you'll learn about this mental function of ours (or "mechanism") known as agency, which goes a long way towards explaining why we are inclined to believe in gods. It's good ammunition in your next debate with a religious believer! Also take note of the list of books on the topic of evolutionary psychology. They're all classics.

DISAGREE:

The original wording I used for that statement was this:

> **STATEMENT**: Our tendency to believe in
> gods is a by-product of the human mind's
> cognitive mechanisms.

Though a more accurate way to state it, that version was a bit erudite, using terms like "by-product" and "cognitive mechanisms." Regardless of the wording, the point is the same: It seems we evolved certain mental traits that cause us to readily believe in gods. You're reading this section because you disagree with that, but before you completely dismiss the idea, allow me to elaborate.

Imagine you're on an "Extreme Eco-Tourism" adventure, hoping to take some photos of the wildlife like jaguars and anacondas. Early one morning, waking before the others, you decide to set out on your own to get some pictures. After wandering for a while, you head back...but your camp isn't where you thought it was. In a panic, you spin around quickly, *Where's the path I was on? Everything looks the same!* Your heart is racing, now. You're lost in the jungle.

Can you imagine that? Wouldn't that be terrifying, especially as you call out for help and no one answers? You take a few hurried steps in one direction, yelling, hoping to wake the others...if you're still in earshot. Now you stand still, listening for any response. And then, something on your left side. A large fern is moving. You scuttle backwards as you turn to face it. Oh, it's just some water dripping onto the fern that caused the movement.

Your ancestors lived in this kind of constant state of "alone in the jungle." To see why that's relevant, let's invent two hypothetical ancestors of ours. They're brothers, living in that very same jungle. One of them jumps at seemingly every bit of movement. He assumes that things like rustling grass or shaking ferns are caused by a "conscious agent." His brave brother, on the other hand, operates on the principle that the movement of most inanimate objects is caused by the wind or falling water. Tell me: Which of these two brothers is more likely to live long enough to have kids?

We are the descendants of people who attributed "agency" to nearly everything, and therefore lived long enough to pass down that trait. Sure, the jumpy brother wastes a lot of energy being startled by unfamiliar sounds, and uncertain movements. But his brave brother who *never* assumed that such things were caused by a conscious agent, well....by the time he determined that it really *was* a jaguar lunging from behind the fern, it was too late. The predator had him.

Thus we have an evolved tendency to attribute agency to things. *That's* what scientists mean by a "cognitive mechanism." And this mechanism, along with many others, primes us to believe in all sorts of supernatural agents including ghosts, spirits...and gods. So, when my Christian friend marvels at a beautiful sunset, saying, "Gosh, the Lord sure created a great show for us!" he is unconsciously attributing agency to the sky.

Compare this:

> "**An intelligent, unknown being is causing that fern to move**!" says the brain of our jumpy ancestor.

...with this:

> "**An intelligent, unknown being is causing that beautiful sunset**!" says the brain of my spiritual friend.

Agency attribution is just one of the many cognitive mechanisms that we've inherited from our ancestors and which contributes to our predilection for all things spiritual. To describe all these mechanisms in any detail would require a long, jargon-heavy chapter, which is beyond the scope of this book (to say nothing of my being beyond my rudimentary grasp of evolutionary psychology.) But there are many excellent books on this very topic. (See the list below.)

Besides, my goal here is only to get you to ponder your own spirituality. It's my hope that you might reconsider the root cause of your spiritual feelings. *Are you certain an intelligent, unknown "sunset creator" exists, or could it be possible that you're inadvertently attributing agency to the sky?*

* * * *

If you've never read a book on the topic of evolutionary psychology, then allow me to make some recommendations. These are just a few of my favorites, listed in no particular order. As I read each of these, nearly every page brings a sense of, "Oh, that makes sense! Now I understand why I do that!" It really is a fascinating topic.

RESOURCES

How the Mind Works by Steven Pinker

Why We Believe What We Believe: Uncovering Our Biological Need for Meaning, Spirituality, and Truth by Andrew Newberg, Mark Robert Waldman

The Red Queen: Sex and the Evolution of Human Nature by Matt Ridley

Why Beautiful People Have More Daughters: From Dating, Shopping, and Praying to Going to War and Becoming a Billionaire-- Two Evolutionary Psychologists Explain Why We Do What We Do by Alan S. Miller and Satoshi Kanazawa

Why We Believe in God(s): A Concise Guide to the Science of Faith by J. Anderson Thomson

The Evolution Of Desire by David M. Buss

The Selfish Gene: by Richard Dawkins

Survival of the Prettiest: The Science of Beauty by Nancy L. Etcoff

STATEMENT: You're either pregnant or you're not.

AGREE:
My high school science teacher used to say this all the time. It was the analogy he used to urge his students to give Yes or No answers to whatever question he just asked the class. Back then, I didn't have the courage to confront him on the validity of his claim. I do now.

Tell me, is a woman pregnant when the baby is only halfway out of her during delivery? If so, then when does she stop being pregnant? Imagine watching a delivery. At what precise moment do you decide, "Okay, *now* she's not pregnant." When the entire baby is out? What about during that time while the umbilical cord is still connecting them, and the baby hasn't taken its first breath? Even though the baby is out, it is still reliant on the mother for oxygen just as when it was inside.

More importantly, when exactly does a woman *begin* to be pregnant? If you say it starts at conception, that doesn't help much. The Merriam-Webster dictionary defines conception as: *the process of becoming pregnant involving fertilization or implantation or both.* So, which do you mean? Fertilization or

23

implantation? How about *as* the sperm is entering the egg, but it's not fully inside yet? Is the woman pregnant then? And what about a fertilized egg that, for whatever reason, never implants in the uterus? If you consider that pregnancy begins with fertilization, then when exactly does it *end* in the case of non-implantation? And for either case, fertilization or implantation, when exactly does pregnancy end in the case of miscarriage? Does it end before the miscarriage process starts? When, exactly?

A survey published in the American Journal of Obstetrics and Gynecology showed that out of 1,000 ob-gyns, 57% felt that pregnancy begins at conception (there's that vague word again), 28% felt it begins at implantation, and the remaining 15% were unsure. (Chung) Can't we admit that pregnancy isn't a clearly focused, black and white issue? It's fuzzy and gray. Here we have medical professionals who can't come to any firm consensus on when pregnancy starts, yet we have lay people boldly declaring their opinion to be fact which they insist all the rest of the world must recognize.

You'd think if God were so adamant about abortion and stem cell research he would've written unambiguously about them in the Bible. Two additional commandments could've cleared up the whole debate:

Commandment #11:
Thou shalt not have any abortions whatsoever.

Commandment #12:
Thou shalt not conduct any stem cell research.

Why are Christians so adamant that pregnancy starts immediately at fertilization when the word itself is nowhere to be found in the Bible? In fact, why are they so adamant that the fetus is even *alive*, when clearly the Bible places little value on the fetus. Read for yourself. Exodus 21:22-24...

"If people are fighting and hit a pregnant
woman and she gives birth prematurely [has
a miscarriage] but there is no serious injury
(to her), the offender must be fined whatever
the woman's husband demands and the court
allows. But if there is serious injury, you are
to take life for life, eye for eye, tooth for
tooth, hand for hand, foot for foot." (*New
International Version*, Exodus 21:22-24)

No need to rely on my interpretation. Here's what the Christian
evangelical professor Bruce Waltke wrote in the magazine
Christianity Today, an evangelical Christian periodical,
founded by Billy Graham:

"God does not regard the fetus as a soul, no
matter how far gestation has progressed. The
Law plainly exacts: 'If a man kills any
human life he will be put to death' (Lev.
24:17). But according to Exodus 21:22–24,
the destruction of the fetus is not a capital
offense… Clearly, then, in contrast to the
mother, the fetus is not reckoned as a soul."

The publishers of *Christianity Today* agreed with professor
Watke, saying, "The Bible definitely pinpoints a difference in
the value of a fetus and an adult." Even the, shall we say,
somewhat conservative Southern Baptist Convention passed a
resolution in 1971 that abortion should be *legal*. The Christian
mindset on abortion only started to change once Jerry Falwell
hit the airwaves with his personal interpretation of the Bible.
(Dudley)

Not that I want to tackle the issue of abortion itself. Not here,
anyway. I just wanted people to reexamine this fundamental
belief about when pregnancy starts. There's so much at stake on
this, surely it's worth a few minutes of your time to think it
through on your own.

DISAGREE:

I'm surprised anyone is reading this 'DISAGREE' section. Religious or not, everyone seems to feel that a woman is indeed either pregnant or not. This idea of *When does pregnancy start and end?* is one of those beliefs that we rarely revisit. That, of course, is the point of this book. I'm encouraging you to reexamine your long held beliefs. I stress that I'm not necessarily trying to get you to change them, but only ask that you be honest with yourself about how you came to hold them. Did you really sit and ponder for yourself about when pregnancy starts and when a fetus becomes an independent, viable human being, or did some minister simply convince you to interpret the Bible the way he does?

Regarding pregnancy, obviously there is a lot at stake. Hospital policies, medical ethics, stem cell research, and of course local and national legislation regarding abortion are all impacted by the exact definition of pregnancy. I certainly don't claim to have the answer. I'm very comfortable saying, "I don't know exactly when pregnancy starts." What scares me—what should scare every rational thinking person—are all the people who view the gray world of pregnancy through the black-and-white glasses of religion.

STATEMENT: Faith is all a person needs to know the truth.

AGREE:
Pretend you're on trial for a murder you didn't commit. It's a case of mistaken identity, and there are mounds of clear and convincing evidence in your favor proving you couldn't have done it. Unfortunately, the actual murderer—the person you resemble—is a beloved local preacher. And all twelve members of the jury have faith he didn't do it.

As your defense lawyer shows proof of your innocence—you were out of the country, your fingerprints don't match those on the weapon, etc—the jury members barely pay attention. Why should they? They have *faith* that you're the murderer, and faith is all a person needs to know the truth. Isn't that what you agreed to? The jury should be *applauded* for their ability to maintain their faith in the light of such otherwise convincing evidence. Thanks to the irrationality of faith, you're off to the electric chair.

Or take this scenario: You go to your dentist for a mild pain in your jaw. Without taking any x-rays, without so much as glancing in your mouth at all, he declares his faith that you

27

need a root canal. As he prepares for the operation, you willingly submit to the procedure. Why wouldn't you? Faith is all anyone needs in order to know the truth, right?

Last example: I've got a used car for sale. It gets 1000 miles per gallon. Don't ask me to prove it, though. It only delivers that kind of mileage to those who have faith in it.

Oh, what's that you say? Sometimes we *can't* rely on faith? Sometimes we need evidence? Hmm, I guess *now* the rule seems to be, "**Faith is all a person needs in order to know the truth**...

...as long as the issue doesn't relate directly to my health or safety or finances."

Are those the only exceptions to the rule? Can you think of any other situations where we shouldn't rely on faith? Take a moment to think about it.

How about your child's education? Imagine you are homeschooling your child and you've hired a Christian math teacher. As the teacher writes on the whiteboard he explains, "Five plus three is umm...*fifty-three*." He turns to you with a wink, "Don't worry. I have faith that's right." Even the most devout and faith-loving Christian would nevertheless fire such a faith-based tutor on the spot. Since we obviously can't teach math based purely on faith, it seems our list of qualifiers to the original statement needs to be amended again: **Faith is all a person needs to know the truth**...

....as long as the issue doesn't relate directly to my health or safety or finances, *or to my child's understanding of math.*

How about your child's understanding of physics? Is it okay if your homeschool physics tutor tells your child that, through faith, we know that the speed of sound is about fifty miles per hour? Meanwhile, your child's chemistry teacher explains, "Hydrochloric acid is harmless. My faith tells me that." And the history teacher explains he has faith that Sylvester Stallone was the first President of the United States.

I'm guessing you wouldn't let any of those slide by either. So now we have: **Faith is all a person needs to know the truth...**

....as long as the issue doesn't relate directly to my health or safety or finances, *or to my child's understanding of math, basic physics, basic chemistry, and history.*

Certainly you can no longer agree with the original blanket statement that faith is all a person needs to know the truth. Therefore, please read the 'DISAGREE' section below.

DISAGREE:

Let's be clear: Faith is unwavering belief without any evidence. Are there other definitions? Certainly. Faith can refer to a set of religious beliefs, as in the *Zoroastrian faith.* And the word faith can be used to mean **a sense of confidence in a successful outcome,** as in, "I have faith our team will win tomorrow." Or, "I have faith I'll get that job." In these kinds of sentences, you don't really ***know*** your team will win, or that you'll get that job. What you mean is, *you feel confident the outcome will be successful.*

Likewise, when you say you have ***faith in God,*** surely you don't mean to say you have confidence that He will have a successful outcome, as if He were participating in some sporting event, do you? "Come on God, you can do it! I have *faith* in you!"

Of course not. When a religious person says he has faith in God, he means he believes unwaveringly, and without the need for any evidence. And the stronger a person holds on to their faith—especially when people around you are trying to convince you that you're wrong—then the more admirable you become. It seems so pure and sweet and childlike. There's something romantic, even, about putting all your trust in another individual.

In reality, faith is a dangerous thing. Faith leads men to fly planes into buildings. Faith leads people to hunt "witches" in Salem. Faith lead Christian Scientists to "treat" 11-year-old Ian Lundmann's diabetes with nothing but prayer, which directly and senselessly resulted in his death. Faith is nothing less than the enemy of reason.

I understand that faith gives people great comfort. I'm sure it feels wonderful to have faith that an invisible super-parent is watching over you and has great plans for you. And it must feel equally wonderful to believe unwaveringly in a magical plane of existence where your mind and memories live on in an eternal family reunion. It's precisely those warm, comforting feelings that I'm up against whenever I ask someone of faith to reexamine one of their beliefs. But there's so much at stake, I have to keep trying.

CRITICAL THINKING LESSONS
FROM CHAPTER 1

Here are the takeaways from Chapter 1:

* It's a good idea to question even your most fundamental beliefs.

* Even brilliant people can believe weird things, and therefore even *they* need to stop once in a while and question their fundamental beliefs.

* Most of our beliefs are established without evidence.

* Faith is belief without evidence. It's irrational and harmful.

PARTING THOUGHT:
Questioning My Own Fundamental Beliefs

It seems that in my early teens I developed all sorts of bizarre and unfounded beliefs. I don't mean paranormal things. These were mostly cultural beliefs. For one, I believed that rap music was terrible, even though I barely heard enough to judge it. I was a hard rock fan. Led Zeppelin, Rush, Van Halen, AC/DC. Am I forgetting anyone? I was so closed minded it's embarrassing. I clung to my belief about rap for years, even as I began to hear more of it and—somewhere in my mind— noticed I actually *enjoyed* some of it. How weird is that? There I was, enjoying the rap I was hearing, and yet because my belief was, "Rap is bad. Rap is for dumb people," I never listened to it. Thankfully, in my thirties I finally began to do an honest appraisal of my beliefs and I noticed the hypocrisy. Am I now a huge rap fan? Not really. But there are times when the mood hits me and I love listening to it.

Another odd belief formed in my teen years? I believed I wasn't good at foreign languages. In college, I actually changed my major because the original one I wanted to pursue required two years of foreign language study. Imagine: Changing your major because you believe "learning a language is too hard." Isn't that strange? I held that particular belief until I was nearly thirty-five. Then, in a major confrontation with my belief, I set out to learn Russian. A few years later I was completely fluent and actually moved to Ukraine to immerse myself in the language.

I had many more odd beliefs left over from those troubled years; beliefs about food, clothing styles, hair styles, drugs, women, and so on. I wish I could express how refreshing it is to let your choices and actions be dictated by reason, not by irrational and foundless beliefs. In this book we only have time for the Big Beliefs, but it could be worth your time to reexamine some of your smaller beliefs, too. (Just don't blame me if you end up moving to Kiev.)

32

CHAPTER 2:
You're Missing Out
on Something Wonderful!

INTRODUCTION:

In the first chapter I wanted to show that there are compelling reasons to reexamine our fundamental beliefs. There is so much at stake for all of us, it's worth the time to closely examine at least some of them. Having established that it's okay to question a particular belief, the next logical questions are: *When* and *why?*

Before reading the first statement of this second chapter, please take a moment to ask yourself if you would ever change one of your beliefs, and why?

STATEMENT: A person's beliefs should *never* change.

AGREE:
You're reading this section because you agree with the statement that **a person's beliefs should never change**.

It's okay to change your answer, by the way. If you're having second thoughts and you now feel there *are* times when someone should change a particular belief, then please jump ahead to the DISAGREE section below.

Hmm....You're still here. You *really* feel that all beliefs should be unchanging. Well, then I have to ask: Have you never changed a single belief of yours in your whole life? For example, perhaps you believed in the Easter bunny when you were a child. Do you still believe in the existence of a magical rabbit that hides painted eggs each Easter, or did you change that belief?

Did you believe in Santa Claus as a child? Depending on your religion, perhaps you didn't. But you know that millions of adults alive today *used to* believe in Santa Claus when they were children, but changed their belief when they got older.

Are you saying they were wrong to change their belief?

I'm really hoping I've changed your mind on this point, and that you now agree: **There are times when a person should change their belief**. In which case, please read the section below.

DISAGREE

You're reading this section because you *disagree* with the statement that **a person's beliefs should never change**. In other words, you feel that in at least some circumstances, a person *should* change his or her belief about something.

That begs the question: **How can a person know if they should reconsider a particular belief?** We each have thousands of beliefs. No one has the time to reexamine each one.

This is the scariest part of the book for most people. Our beliefs are such a major part of who we are. If we change one of our Big Beliefs we risk losing friendships, relationships, even our *marriage*. And though your own family may not reject you, they'd certainly be deeply hurt and upset.

You're probably thinking: "Why should I consider changing my belief? It's working great for me. Most of the people around me seem to believe the same thing, and it's working great for them, too. My beliefs certainly aren't *hurting* me. In fact, they give me comfort." I understand that. Believe me, I do. But here's why you should reexamine your core beliefs:

You're missing out on something wonderful.

For example (and this is true): I have a friend who has never watched the movie TITANIC because he believes it's a bad movie. For some reason, he believes he wouldn't like it. He believes this even though I, and most of his other friends, tell him it's great and that he'd really like it. He clings to his belief despite the overwhelmingly positive reviews TITANIC received. Isn't that sad? It is a great movie, and he's missing out on something wonderful simply due to a bizarre belief.

I have another friend who believes the Apollo moon landing of 1969 was a government hoax, and that men have never walked on the moon. Again: Isn't that sad? What a fantastic thing those men did back then. How sad to not share in the awe of what they accomplished! He is missing out in rejoicing over one of mankind's greatest achievements.

And I have another friend who, well....he believes in an invisible sky fairy. Without getting into the specifics of which religion he adheres to, suffice it to say he believes that this magical sky fairy created the universe, including the earth and all life on it. His sole source of "evidence" for this belief is a magical book he believes was dictated by his sky fairy. (And please don't assume I'm referring to *your* particular religion. There are *lots* of religions that fit the above description.)

So, what is this third friend of mine missing out on? Oh, I don't know, how about: *EVERYTHING!* You think TITANIC is an interesting story that shouldn't be missed? Well, the story of evolution is a multi-billion-year-long *epic!* It's a mind-boggling tale, and one that all of us are a part of. And how about the Big Bang? It's another gripping story about the birth of galaxies and the deaths of stars, and again, you and I are a part of it. And yet, despite literal and figurative *mountains* of evidence in support of the Big Bang and evolution, my friend chooses instead to believe that his invisible, undetectable, magic sky fairy simply willed everything into existence.

To me, that is crushingly sad.

This religious friend of mine has a three year old son. If the child one day asks, "Daddy, how do cars work?" I wonder if my friend will lift the hood and explain (as best as he himself understands it), that, "A bit of gasoline is ignited inside a piston, and the explosion moves the piston, which eventually helps turn the wheels." Or will he just say, "The car fairy makes them go."

How cruel, to stifle the natural wonder and curiosity that children have about the world around them by explaining, "God did it." Don't you agree?

STATEMENT: The Great Pyramid of Giza was built by ancient Egyptians (not by extraterrestrials).

AGREE:

Like most people, you believe the pyramids were built by humans, namely, the ancient Egyptians. (And not by extraterrestrials.) Although you and I are both in agreement, it's worth examining how the non-believers (that is, those who feel the pyramids were built by aliens) came to their conclusion.

Usually they jump to such a conclusion as a result of some hard-to-answer questions, like: "How could they lift such massive stones?" or "How could they build it so precisely?" Those are indeed great questions worthy of research and debate. But such questions in no way prove (or even suggest) extraterrestrial involvement. There are actually a few competing explanations for how the Egyptians built the pyramids: Was it slave labor or skilled craftsman?, ramps versus levers?, and so on.

Here's the key point. **It's great to ask tough questions**, like: "What caused the Big Bang? How do inanimate molecules suddenly become alive? How did the ancient Egyptians lift such enormous stones?" and so on. But when questions go unanswered or you disagree with a particular answer, that does *not* mean that *your* particular explanation is correct. Think of it this way:

Pretend we've just watched a show by a professional illusionist and now we're discussing his tricks. We're stumped, though, by one illusion where he made a car float and then disappear. So one friend says, "Aliens must have levitated it and then beamed it away."

Just because we can't fully explain some process, be it vanishing cars or the construction of ancient pyramids, it doesn't mean that all possible explanations are equally valid, and it certainly doesn't mean your favored explanation wins by default. If you want to convince me that aliens were involved in building the pyramids, you'll need incontrovertible physical evidence.

DISAGREE:
If you honestly feel that the pyramids were not built solely by the ancient Egyptians, I'm going to assume you therefore believe that aliens must have been involved. But how do you know it was *aliens* and not the technologically advanced inhabitants from Atlantis? You have to admit that you made the jump to "aliens did it" because you don't fully understand how the Egyptians alone could have done it, not because you have some physical evidence of their involvement, like an alien corpse, or one of their spaceships.

To put it another way, even if we all agree the Egyptians couldn't have built the pyramids, you still need to have concrete evidence to support your particular theory that extraterrestrials did it. Because I repeat: **How do you know the pyramids weren't built by the technologically advanced inhabitants from Atlantis?** Or by future inhabitants of Earth who've learned how to time travel and decided to go back and help build the pyramids? Perhaps the Egyptian sun god Ra actually exists and he's the one who helped them. Maybe there was a family of mind-bogglingly strong Egyptians who did the bulk of the lifting.

The funny thing is, you're probably shaking your head, "Inhabitants from Atlantis? The sun god Ra? That's *ridiculous!"*

I see. So, inhabitants of Atlantis is a crazy idea, but incredibly advanced aliens traveling to Earth to help the ancient Egyptians build pyramids *not out of the super-advanced materials they brought with them* but instead out of rock which just happened to be located in nearby quarries, and helping to build them in such a way that the pyramids appear exactly as if the ancient Egyptians themselves had built them unaided....**that idea is *not* crazy?**

You're obviously smart and inquisitive. I know because you're asking some very intelligent questions about how the pyramids came to be. So why not keep going? *Why not ask those same kinds of insightful questions about your own theory?*

* If your aliens are so advanced, then *why are the pyramids so poorly constructed?* They're falling apart; between the stones there are huge gaps filled with sloppy mortar; the exterior casings are nearly completely gone, and so on. Your aliens have the technology to traverse trillions of miles of space but they can't find a stronger adhesive to prevent looters from stealing the granite exterior?

* And why did they build them out of local limestone? These aliens had spaceships at their disposal. Even if they insisted on using only earthly materials (and not using their own ultra-advanced materials because evidently they didn't want anyone to think aliens had visited Earth) *why not gather up all the gold and diamonds on our planet and make a truly spectacular pyramid?*

* Proponents of the "ancient astronaut" theory love to proclaim that the Pyramid of Giza is aligned to about one-tenth of a degree of due north. I'm sorry guys, but that's not proof of your theory but instead the death-blow of it! Again, these are ultra-advanced extraterrestrials: *How could they make such a huge gaffe?* A one tenth of a percent miscalculation is a gigantic number when dealing with interplanetary (to say nothing of interstellar) space. Our own puny little NASA scientists can calculate not just a simple north-south orientation but complex orbits and trajectories with vastly more precision than that. A one-tenth of a percent mistake is laughable for an extraterrestrial being...

...though it's quite impressive for the ancient Egyptians.

As I mentioned in the "AGREE" section above, it's great to ask tough questions (like how did the Egyptians align the pyramids and move all that rock?) But my inability to answer something does not make your particular theory win by default. And again, the key point in this section is: If you're asking those kinds of intelligent questions about the consensus theory then *why not keep going and ask those same kinds of intelligent questions about your own theory?*

And finally, remember one of the main messages of this book: **You're missing out on something wonderful**. The pyramids represent one of the pinnacles of human achievement. Take pride in what those surprisingly clever, resourceful and industrious Egyptians did. Find inspiration in it. Because as cool as your alien fantasy is, you're missing out on the truly fantastic heights that humans are capable of.

Just reading the statement below makes me sick to my stomach. I'm passionate about this one, so bear with me if I vent.

STATEMENT: The purported moon landing in 1969 was actually a hoax perpetrated by the U.S. government.

AGREE:
We've all heard this one, but I never thought anyone except a few crackpots (the kind who live in trailers in the middle of the desert) actually believed it. That is, until a *non*-crackpot friend of mine mentioned his own belief in it about ten years ago. Again, this isn't a debunking book. Every point of contention that the "moon hoax" believers have ever raised has been thoroughly explained by a variety of experts. Type in "moon hoax debunk" into your favorite search engine and take your pick from the long list of results. (My favorite is: http://www.clavius.org/) Remember our primary objective here: **I'd like you to reexamine your belief.** Allow me to do so with a series of questions:

How did you first come to believe the moon landing was faked?
Determining how and when we came to any particular belief can be very enlightening. So tell me, were you on the movie set where Neil Armstrong and Buzz Aldrin filmed their moon walking scenes? Were you working on the movie in post-production and that's where you saw suspicious outtakes like Armstrong with his helmet off, still on the moon soundstage, taking a cigarette break?

Do you have any direct observations at all that it was faked?
No, you don't. What you have are questions. Lots and lots of questions. More on those in a moment, but I want to get to the bottom of how you first came to your moon hoax conclusion.

Who planted the seed of doubt?
I'm guessing it was some conspiracy-themed website, or a fellow conspiracy-minded friend. In my friend's case, his older brother is a professional photographer who feels there are "discrepancies" in the moon photographs. Clearly, his brother is the source of doubt.

What does that tell us? Well, if you're honest, you'll at least admit that you have a tendency to rely on false authority, or fringe authority. It's not like the website you got your "information" from was any of the following: ESA (European Space Agency), NSF (National Science Foundation), or the AAAS (American Association for the Advancement of Science). And I'm sorry, Thomas (not my friend's real name), but your brother is not an expert on lunar photography, is he?

Why not take NASA to court?
You have all this "proof" about impossible photos with multiple light sources, the impossible angles of shadows, and "impossible video of a moving flag on the moon." It's incontrovertible, isn't it? So it should be an absolute cakewalk for you in court. Lawyers love big cases, especially the easy ones, so you shouldn't have any problem finding an eager lawyer. So what's stopping you?

Have you honestly taken the time to carefully research the answers to your moon landing questions?
Lunar physics is a tricky thing, so I understand why you might have trouble comprehending certain aspects of the moon landings. I'm certainly no expert, but I carefully read the explanations from certified science experts. When you ask questions like, *Why is the flag waving if there's no atmosphere on the moon?* or *Why are there no stars visible in the photos?* and so on, what you're really saying is, "I'm not smart enough to understand how the flag could move if there's no air on the moon." "I'm not smart enough to understand why dim stars wouldn't show up in the background of a photo, when taking a picture of the bright lunar surface." (I have to ask: Do you really think NASA would overlook that? *Oh shoot! We forgot to add stars!*)

In this case, you guys resemble the religious fundamentalists who use this same **argument from ignorance** (as this fallacy is called) to pose questions about, say, evolution. "I'm not smart enough to imagine how the flagellum could have evolved. I'm not smart enough to imagine how the eye evolved." Instead of proclaiming your ignorance, you should be reading actual science books which explain the things you don't understand.

Why don't you ever ask similar questions of your own theory?
It's good to ask questions like the ones above, but you need to do so fairly and consistently. The questions you moon-hoaxers ask are always aimed at casting doubt on the mainstream explanation. Why do you not take aim at your own theory?

Imagine if your moon-hoax view somehow becomes the mainstream view. (Which is actually conceivable, thanks to irresponsible network programming like FOX airing a moon-conspiracy show called: *Did We Land On The Moon?*) If your view was mainstream, you'd tear into it: There were hundreds of thousands of people who worked on the Apollo mission during those ten years or so. Anyone who comes forward with proof that it was a hoax would become rich. *How do you keep them all quiet?* The missions returned 800 lbs of moon rocks, which have been studied by independent geologists for decades. *Are they all being forced to conceal the truth as well?* And one of the toughest for moon hoaxers to explain away: *How do you explain the existence of laser reflectors on the moon?*

You don't ask these kinds of questions about your own theory because you're not interested in the truth. (If you were, you would read books on science.) Instead, you're interested in feeling like you have secret knowledge that few others do.

Are the current Mars probes also hoaxes?
What I find most telling is that none of you are crying "Hoax!" about the current Mars rovers. But why not? As impressive as the moon landings were, they absolutely pale in comparison to the scope and complexity of what NASA has achieved with the Mars rovers, as well as its numerous other incredible probes, such as the one that set down on Saturn's moon Titan. Such a hoax would be much easier to pull off because you just need to film a rover, and not live humans.

Don't you realize what you're missing out on?
It might feel cool thinking you're in on some secret that no one else knows about, but it's not. It's sad, because you're missing out on something wonderful. These astronauts (some of whom gave their lives to the cause) did a spectacular thing! That series of moon landings is arguably the greatest achievement in the history of mankind.

You think you're standing high up in your Tower of Superior Knowledge, when you're actually deep down in the Sad Pit of Close-Minded Ignorance. But I'm there at the edge of your pit, reaching out my hand. If some of the above questions have gotten you to reconsider your moon hoax belief, then I encourage you to come up and join us here in the land of Rational Thought. We'd love to have you. (Again, it's great to hang out with people who ask smart questions, but you need to listen carefully to the answers.) But before I pull you out, I insist you do one thing: You have to apologize, if only silently, to those who **dedicated** their lives, to those who **risked** their lives, and to those who **lost** their lives to make it happen. And then thank them. *To the Apollo team members: I used to think what you did back then was a hoax, but I've come to realize we really **did** land on the moon. I apologize for the pain my unfounded accusations may have caused. And thank you for helping mankind take that first giant leap towards the stars.*

DISAGREE:

It does make you wonder how any intelligent person could fall for such nonsense, but as I said, I have a good friend who believes the moon landings were all a hoax. Having met other conspiratorial-minded people, I've noticed certain patterns. One is, they tend to subscribe to other conspiracies, too. So, if someone believes the moon hoax theory, it's likely they adhere to the 9/11 conspiracy as well, or to one of the many "Barrack Obama" conspiracies. They also seem to equate "mainstream" with "always wrong" or "worthy of perpetual distrust."

Ultimately, **everything we do is either to increase pleasure or minimize pain.** Viewing their odd beliefs that way, things start to make sense. A moon hoaxer must get more pleasure out of considering himself an elite member of those who "know the real truth." Meanwhile, despite mounting evidence contrary to their belief—despite all the rational answers to their conspiracy questions—they cling tightly to their claim. Why? Because they're avoiding the pain of admitting they're wrong.

Remember, this isn't some little mistake they'd have to admit to. This is a **gigantic** one, representing years or even decades of investment in an incorrect theory. Imagine admitting that some story you had convinced yourself of and preached to others as being the "real truth" has actually been a bizarre delusion. That's an incredibly painful thing to admit.

In the section below, I'm imagining my reader to be at least somewhat religious, and therefore troubled to some degree by the scientific explanation for life's origin.

STATEMENT: The presence of highly specified complexity in any object or system always indicates an intelligent designer.

AGREE:
The classic creationist (a.k.a. "intelligent design") argument features a pocketwatch. They argue that if you're walking through the forest and happen upon a pocketwatch, you'd feel certain the object was created by an intelligent designer. It is not only complex, but was clearly created for a particular purpose. And it certainly couldn't have come about by chance. By analogy, they then reason that the human eye or the human brain must also have been designed, since they are even more complex, and also have a clear purpose.

What I find ironic about the "pocketwatch in the woods" argument is that it ignores the very fact that *pocketwatches also evolved!* Do you think the very first time-keeping device looked anything like a Charles Frodsham English pocketwatch

with a repeating minute, a perpetual calendar, a minute &
seconds chronograph, and moon phases?

It didn't. That pocketwatch you stumbled upon in your
creationist analogy evolved over many, many generations.
Take a look at the distant ancestors of that watch and you'll see
a very crude, clunky and awfully inaccurate clock. Then,
building on that first device, there followed generation upon
generation of small changes and gradual improvements—
including many detrimental changes and experiments
(mutations) that didn't get passed on (evolutionary dead ends)
—and that's how that watch came to be. The clocks that kept
the best time (the ones that were the most "fit") were quite
literally *reproduced*. For my money, the "pocketwatch in the
woods" argument is a great analogy for evolution...

...except for that bit about "chance." There *was*, after all, a
designer for all of those clocks and watches. They didn't
happen by chance. But this analogy we've been working with
—that of a watch—is particularly apt, because the variable here
is *time*.

I explain this in detail in the section below, but I can't just ask
you to jump to that section. I first need to get you to disagree
with the original statement, namely: The presence of highly
specified complexity in any object or system *always* indicates
an intelligent designer.

The key word is *always*. Does purposeful complexity *usually*
indicate a designer? Yes. But *always?* No. Therefore, all I need
is one good example of a complex design occurring naturally,
and I can direct you to the "DISAGREE" section.

Before continuing, though, we need to define our terms. Let's
define "design" as **the process of creation by an intentional
agent.** So, is there any complex structure in existence that we
can all agree occurred naturally? I think so. There are quite a
few, as it turns out.

Take a watershed, for example. When it rains, the tiniest rivulets of rainwater join up with bigger tributaries, which themselves join the trunk of the main river itself. This complex branching structure is ideal for maximizing the flow of water from a broad area (the countryside where it rained) to a point (the mouth of a river). You'd be hard pressed to improve it. So, was it designed by an intentional agent, or did such a clever branching structure happen through totally natural causes? Ask yourself: Is God really out there with his magical invisible shovel, digging the trenches and channels for the water to flow through every time it rains? Or does the branching structure occur naturally and spontaneously because *it's simply the easiest way for things to flow?*

If you think about it, that branching structure is ubiquitous in nature, isn't it? It's important to think this through on your own, so please take a minute to think of other places in nature where we see flow being maximized through a branching process.

Did you think of some? Here's one: **Lightning** uses a branching structure to maximize the flow of electrical current from the broad area of a cloud to a point on the ground.

What about a **snowflake**? A snowflake's branched shape arises to maximize the flow of heat as a drop of water vapor freezes.

The lung maximizes the flow of oxygen into, and carbon dioxide out of, the body from a broad area (your lungs) to a point (your trachea).

How about a tree? **A tree** uses the branching structure twice: Once in its roots, to maximize the flow of water and nutrients from the broad area of the ground, up into the trunk. And again the tree uses branches to maximize the flow of water back into the atmosphere.

You might argue that these last two—the tree and the lung—are evidence of God's immaculate design. But the first three: watersheds and lightning and snowflakes? I again must ask if

He is really out there with an invisible shovel, digging little tributaries every time it rains? Is He guiding the flow of electricity in every lightning bolt? And is He really spending his time laboring over every single snowflake?

Let's agree that He's not. Because the classic branching structure that we see time and again in nature, although it *seems* designed by an intentional agent, is actually a natural process that happens any time things flow. (To learn more about this, please read the book, <u>Design in Nature: How the Constructal Law Governs Evolution in Biology, Physics, Technology and Social Organization</u> by Adrian Bejan and J. Peder Zane.)

Is the branching structure of a tree or a lightning bolt or a snowflake anywhere near the complexity of an eye or the human brain? No. But remember: I mentioned that the critical variable here is *time*. Rivers and lightning bolts and snowflakes come into existence quite quickly. Eyes and brains have been developing very, very, very slowly. (And I can't help pointing out that both the eye and the brain utilize branching structures themselves: Neurons and their dendrites are branched to maximize the flow of electrical signals in the brain. And the eye, like the rest of the body, contains veins and capillaries— the classic branching system—to maximize the flow of blood.)

I hope we now agree: Sometimes, the complexity we see in an object or system does *not* necessitate an intelligent designer. Assuming you're in agreement on this, please read the section below to see the importance of time as it regards apparent design in more complex things like the brain.

DISAGREE:

You're reading this section because you *disagree* with the statement that complexity always indicates an intelligent designer. However, if you lean towards accepting some version of an intelligent design theory to explain the complexity of living things, then you're probably only willing to grant me mildly complex things like rivers and lightning bolts and snowflakes. I suspect you'd still argue that the immense complexity of living things still posits a designer.

As I mentioned in the section above, the key issue is time. Sure, great complexity usually indicates an intelligent designer, but not *always*. **It depends on how long it took for the thing in question to come into being.** Something as complex as your laptop computer, Shakespeare's "Hamlet," or my mother's Chocolate Angel Pie all must have been intelligently designed because *they came about relatively quickly.*

But what about something as complex as the human brain?

Well, it's not like a human brain just popped into existence on your kitchen table, did it? There's no need to postulate an intelligent designer here because the human brain has been three billion years or so in the making. There is an *unbroken chain of custody*, if you will, going inexorably back those three billion, six hundred million years.

Think of it this way:

Imagine a camera that can be passed backwards through time, child to mother, down the long chain of life. I take a photograph of myself, and then pass the camera to my mom. She takes a photo of herself, and passes it to her mother, and so on....and on...and on. A hundred years back...a thousand...a million years....ten million...a hundred million...a *billion*...two billion...*three and a half billion years of photos*, one every generation. If you stacked all those photos one on top of the other, they would reach the top of a 1,000 foot skyscraper...

51

...5000 times! Imagine that. Five thousand stacks of photographs, each stack as tall as a Manhattan skyscraper. ***That's*** your family photo album. [Here's the math: Let's figure that one generation lasts a year. Obviously, with humans a generation is actually 15 - 20 yrs. Meanwhile, on the opposite end, bacteria have generations measured in ***hours.*** But for purposes of this analogy, one year per generation is a fair, if incredibly rough, estimate.]

Getting back to our 5,000 skyscraper stacks of photos: It may not be pleasant to consider, but as you skim through those photos, looking at your great-great-great grandmother, and your great-great-great...plus ***5,000 more greats****....-great-grandmother, you'd likely be shocked to see who you direct relatives were. **At some point, you wouldn't consider your own relatives to be human**. That's bizarre when you ponder it. Bizarre, but no less true.

And that's just the start of the weirdness. At some point, your distant relatives no longer have a shred of "human-ness" about them. Let's go to the 10th stack or so and pull out a photo at random. This would be a photo of your "ultra-grandmother" (that is, a "great-grandmother" times 10 million, give or take). Once again, it would be pretty distressing: This direct relative of yours would resemble a shrew or mouse. But we're only getting started. ***There are still 4,990 stacks of photos to go.***

An important thing to note is how each photo looks very much like the ones right before it and right after it. That is to say, grandmother looks like mother, and mother looks like child. But think of these photos as being frames in a film. Of course the frames that are very near each other will look nearly identical; it's the tiny changes that make movies seamless. But choose two frames that are a few feet apart on the reel and you'll easily see that changes are taking place. Similarly, to notice the accumulation of tiny changes in our evolution, you need to compare photos that are further apart.

If each towering stack of these family photos were a yard apart, it'd take you about an hour to walk along them; a three mile hike to get to that most distant stack. And if we take a look at the photo on top of that final stack, the very oldest picture in *all of our* family albums, you'd see...

...well...I can't quite say what you'd see. I can postulate, as others have. We'd have to zoom in a *lot* to see it, because it was tiny. (I have to call it "it," by the way. This is long before our relatives discovered the benefits of reproducing sexually.) Most likely this ultimate common ancestor of all life on Earth—this alpha-parent—resembled a twisted ladder.

Pretend we had a *movie* of this alpha-parent of ours, instead of just a photo. A movie, in fact, of the very moment when that twisted ladder could be considered alive. Perhaps what we might see is that ladder split apart down the middle of the rungs, and then smaller molecules would fill in the gaps on both halves. And then there were two. It's that simple. The starting point of life on Earth. Add in natural selection (and the much overlooked sexual selection), allow a few billion years to pass, and we arrive at the incredibly complex human brain without any help (or need) for a designer.

Can I say for sure that some intelligent being didn't put that original self-replicating molecule together? Well, no, I can't. I suppose an advanced being from another planet could have done that. But there's no reason to postulate one. There are enough planets and enough time; a self-replicating molecule was bound to turn up somewhere. It turned up here. No big deal.

The goal of this book is to get people to reexamine some of their core beliefs. And this one, the origin of life on Earth, is one of the most fundamental beliefs that I'm hoping you'll reconsider. Assuming you're somewhat religious, though, I suspect this is where you and I will part ways. It's just too uncomfortable—too *daunting*—for most people to change a fundamental belief. You'd have to face your parents, your spouse, your friends, your co-workers....you'd have to admit to all of them that you believe differently now. If you let it be known that you accept evolution as fact, they'll confront you with all the same questions that you once used against other "scientific types."

Remember, though, the theme of this chapter: **By holding on to your religion's fairy tale explanations for the universe, you're missing out on the awe-inspiring intricacy of reality.** Sure, it's a bit unsettling to think that my ultra-grandmother was a forest rodent, but it's the truth. The story of how you and everything around you came to be is a 13 billion year old mystery, worthy of real answers (no matter how disturbing), and not to be brushed off with some pacifying fairy tale.

STATEMENT: You can't prove a negative.

AGREE:
When a believer is backed into a corner and has extinguished his supply of "proof" for his claim, this rusty old shield is one of his final defenses against the relentless attack of reason: "You can't disprove what I believe in!" he declares, hoping to retreat to a stalemate position. "You can't prove that aliens *didn't* build the pyramids! You can't prove that ghosts *don't* exist! You can't prove Jesus *didn't* walk on water!"

I understand why believers like to rely on that statement, but what dismays me is how often this is quoted in the *skeptical* community. Michael Shermer, James Randi, and so on...you guys are my heroes! Would that I could write or argue as lucidly as any of you, or that I were as well informed as you. But may I humbly submit: You damn well *can* prove a negative, at least as far as you can prove anything at all.

Think about it: Isn't it self-evident that something cannot be true and false at the same time? For example, if it's true that apples are a fruit, then the contradictory statement, "Apples are *not* fruit," cannot also be true. Assuming we're all in agreement

55

on this premise, then that's all you need to prove a negative. If I can prove that apples are indeed a fruit, then I can just as easily prove the negative version: "**It is false that apples are _not_ a fruit**."

That seems to me to be simple, self-evident logic. If you now agree that it *is* possible to prove certain negatives, then please read the 'DISAGREE' discussion below.

DISAGREE:
"You can't prove a negative" seems to be a kind of folk logic. I get the feeling some respected author first made the claim, and all his followers echoed it in their own writings, thereby propagating the myth. But if you ask professional logicians, they'll invoke the same reasoning I used above, referring to it as the **Law of Non-contradiction**.

I think what people actually mean when they say "You can't prove a negative," is that you can't prove **universal nonexistence**. That is, I can't prove that a particular thing doesn't exist *somewhere* in the universe. For example, I can't prove that huge, dinosaur-like creatures that have the ability to fly and exhale fire don't exist *somewhere* in the universe. But I can prove *local nonexistence*. That is, as long as we agree on the definition of dragons, I can prove that there are no dragons currently in my garage.

But all this talk about proving negatives is irrelevant, really. It's not up to me to disprove your claim. If those are the rules of debate, I can keep you plenty busy trying to disprove my own fantastical claims. Be my guest: Prove there aren't advanced beings living in a secret city known as Atlantis somewhere in the bedrock of the ocean floor. Prove that Corthu the Almighty didn't create all the other gods.

What's really at issue here is the concept of falsification and the characteristics of a scientific claim. We'll discuss that important topic in Chapter 6, but as we finish this chapter, let's take a moment to ask ourselves this: Instead of demanding that I disprove your claim, perhaps you should ask yourself **What reason there is to believe it?** As you'll see, this is the topic of the next chapter.

CRITICAL THINKING LESSONS
FROM CHAPTER 2

Here are the takeaways from Chapter 2:

* By maintain your irrational beliefs, you're missing out on something wonderful.

* Just because we can't fully explain some process, that doesn't mean your favored explanation wins by default.

* It's good to ask tough questions about the mainstream theory, but why not ask those same kinds of insightful questions about your own theory?

* By holding on to your religion's fairy tale explanations for the universe, you're missing out on the awe-inspiring intricacy of reality.

* Everything we do is either to increase pleasure or minimize pain. This explains the stubborn belief in so many irrational things.

PARTING THOUGHT:
The Things I was Missing Out On

If you read the "Parting Thought" essay at the end of Chapter 1, you'll recall how I was lamenting the fact that I formed so many unfounded beliefs in my teen years: "Rap music is bad," or, "I can't learn a foreign language," and so on. The unspoken tragedy was all the amazing experiences I'd been missing out on. To put it into economic terms: **Every belief comes with an opportunity cost.** If you believe rap music is bad without ever having really listened to it, you're missing out on some serious lyrical prowess. These guys are skillful urban poets, and they're anything but dumb. The interior rhyme schemes, the mastery of rhythm and especially polyrhythm in their delivery, to say nothing of their ability to *improvise* similar kinds of rhymes...there is a lot of noteworthy talent in the rap world. All of which I'd been missing out on thanks to a lousy, close-minded belief.

And I get goosebumps wondering where I'd be right now if I hadn't confronted my belief that I couldn't learn a foreign language. For one thing, I wouldn't have met my amazing Ukrainian wife. I would never have known her loving family, nor any of the lifelong friends I made in Ukraine and Russia over the years. All the incredible things I experienced in both those countries, none of them would have happened if I hadn't reexamined that one debilitating belief.

Tell me, do you have any personal beliefs like that?

CHAPTER 3:
What Reason Is There
To Believe That?

INTRODUCTION:

The statements we'll be discussing in Chapter 3 are a grab bag
of claims all chosen because they make a rational person ask,
"What reason is there to believe that?" I'm sure that you're
familiar with some of the claims in this chapter, but a few will
likely be new to you. As always, please read each statement
carefully and ponder it for a while before deciding whether you
agree or disagree. And of course, I hope you'll read both sides
of the discussion to get a better picture of the argument.

STATEMENT: There is a box buried somewhere in the forests of Austria, inside of which is the manuscript for a 10th symphony which Beethoven secretly wrote.

AGREE:
I'm sure you don't actually agree that Beethoven wrote a 10th symphony and that it lies buried in the ground somewhere. You're just reading this section because I asked you to read both points of view throughout the book. But I love that concept of a closed box, so allow me to expand on it for a moment. As it turns out I happen to have a large wooden box, locked with a padlock, stored in our attic. And to believers I would ask: "Do you believe there's a baseball glove in there, signed by Babe Ruth?" I'm not saying there *is* a baseball glove in there. Maybe there is, maybe there isn't. I'm asking *if you believe* there's one.

Pay close attention to how you weigh your decision. Hopefully, it matches the process described below...

DISAGREE:

That'd be amazing if there really were a box buried somewhere which contained the manuscript for a 10th symphony from Beethoven. It's certainly not impossible. But as with all sorts of amazing claims, from "Michael Jackson faked his death and is still alive," to "The U.S. government secretly records and analyzes every single phone call made in the United States," we need to ask this fundamental question: **"What *reason* is there to believe that?"** I'm not against holding any belief, but I do need a reason to consider it to be true. That's a key point of this book, and of the skeptical mindset: Get into the habit of asking:

> * Which unexplained phenomenon is your hypothesis shedding light on?

Another way to phrase that is...

> * What mystery are you finally solving with your proposition?

Was Beethoven known to have written a 10th symphony which mysteriously disappeared? No. Beethoven never published a 10th symphony, and his collection of notebooks only have fragments of ideas he seemed to be working on. Still, it's possible he wrote the 10th symphony secretly, perhaps at the same time he was writing the 9th. But what evidence is there for that? For example, is there any mention of a 10th symphony in the writings of those who knew Beethoven in his final years?

The fact is, the consensus of experts is that the 9th was Beethoven's last symphony. Why should I even consider there to be a 10th? If you can get me past that, I'll then need to know why I should believe it's buried in a box in an Austrian forest. So, with this claim about a 10th symphony, there *is no unexplained phenomenon* that your hypothesis is shedding light on, and *there is no mystery* that you're solving with your proposition.

The same is true for so many incredible claims. You want me to believe that Michael Jackson faked his death and is still alive? (This used to be claimed about Elvis, but he would've turned seventy-seven in 2012, so the myth is finally dying, and being resurrected in claims about Michael Jackson and rapper Tupac Shakur, among others.) What are the unexplained phenomena? Are there documented photos or videos of Michael, clearly taken after his funeral?

Keep that cornerstone of skeptical inquiry ("What *reason* is there to believe that?") in mind as you read the next statement. It's a real-world example that applies to anyone who uses a cellphone.

STATEMENT: Mobile phone usage can cause brain tumors.

AGREE:
Such a correlation was first proposed in the late 1990's and it's certainly a scary thought if true. But remember the skeptical mindset that we established in the last section: **What reason is there to believe that?**

> * Which unexplained phenomenon is your hypothesis shedding light on?

Are lots of cellphone users suddenly being diagnosed with brain tumors? Are brain tumors more common in people who spend lots of time on their cellphone?

The answers are No, and No.

> * What mystery are you finally solving with your proposition?

Was there a large group of people suddenly getting brain tumors, beyond what we'd expect statistically? Has there been some outbreak of brain tumors that mystified doctors?

Again: No, and No.

Simply put, ***there is no reason to believe that cellphone usage causes brain tumors.*** True, cellphone antennae do emit a form of electromagnetic radiation known as radio frequency energy, but all evidence shows that this energy does not cause DNA damage in cells. Furthermore, radio frequency energy has not been found to cause cancer in animals or to enhance the cancer-causing effects of known carcinogens. (National Cancer Institute)

DISAGREE:

There have been some epidemiological studies called "case-control studies" which basically looked for any correlation between cellphone use and brain tumors in large groups of people. The results were inconclusive, and any correlation they might have found is just that: a ***correlation***, not a cause. It could be that people who use cellphones the most tend to do ***some other activity*** which infrequent cellphone users rarely do, and it's this other activity which ***does*** directly cause brain tumors.

For example: Let's imagine that women who love to dance are more likely to get lung cancer. Does that mean dancing ***causes*** lung cancer? No. But women who love to dance tend to smoke more than those who never dance. It's the smoking that causes the cancer, not the dancing. (I emphasize that I made up this dancing example merely as an analogy. No such correlation exists, as far as I know.)

All that being said, if you use your cellphone a lot, it's probably a good idea to use an ear-piece. This will keep the antennae safely away from your brain. And more pressing: An ear piece will allow you to keep both hands on the wheel when driving.

STATEMENT: Human beings have a soul which survives the death of the body.

AGREE:
I can't help but ask: ***What exactly is a soul?*** If neither party knows what it is that they're discussing, then there can be no meaningful exchange of information. We might as well discuss whether plorgs exist. But I'm not writing this to debunk claims for the existence of the soul. I only hope you might reexamine your belief in it. So for the sake of this discussion, let's use the dictionary definition: *The soul is the immaterial part of a human being, which is regarded as immortal.* You might not agree on the exact wording (some religions feel animals have souls, others don't, and so on) but the general consensus seems to be:

- The soul is spiritual in nature (as opposed to physical).

- The soul survives the body after death.

What I'd like to know is: **What reason is there to believe that?**

Believers often accuse rational thinkers of **not wanting to believe** in things like God or the soul. To which I respond, *Are you kidding me?* There's nothing I'd like more than to know I'll see my wife and family again after I die, and that all my thoughts and memories will stay with me after death. Who wouldn't want that to be true? But **wanting** something to be true doesn't **make it** true.

So I ask again: What reason is there to believe in the existence of a soul? *What observed yet unexplained phenomenon does your theory cast light on?* The following example might help explain what I'm searching for:

Let's travel back in time to the year 1660. Your name is Edme Mariotte and you're a French physicist who has discovered something truly bizarre: *You can make things disappear.* Heart racing, you grab a small coin from your desk and hurry to the palace to demonstrate your discovery to the royal court. "Keep your gaze fixed right here," you tell a courtier as you point to a button on your jacket. "You see the coin now?" you ask, as you hold it near your jacket button.

"Yes."

"And now?" You slowly move the coin about one foot to the side and suddenly...

"It disappeared!" screams the courtier. "How did you do that?"

Edme Mariotte was the first to discover this flaw in the human visual system. And back then, it was a truly bizarre phenomenon that needed an explanation. Can you imagine postulating a theory to explain it? "My guess is, there's a hole in the eye somewhere. When light falls on this hole, our brains must fill in the missing information by extrapolating from the surrounding visual data." As crazy as that theory must've sounded back then, *there was a reason to believe it*. It was a good explanation for the unexplained phenomenon of the blindspot.

To lay out the bare parts, we have:

Phenomenon: Things disappearing from view

Postulation: Blindspot

With the soul, we have...

Phenomenon: *???*

Postulation: Soul

There is no unexplained phenomenon for which the postulation of a soul becomes necessary.

I imagine believers will list all sorts of things that they feel are the unexplained phenomenon which necessitate the existence of a soul: *How do you explain ghosts?* or *Why can some people see auras?* and so on. Those aren't real phenomena, they're imaginary. The human blindspot is a real, universal phenomenon and I can easily demonstrate it to you. Can anyone demonstrate ghosts or auras to me?

The idea that we each have a soul which endures after our death is surely one of the most fundamental beliefs this book will broach, so I won't be holding my breath, hoping you reexamine it. Personally, of all the irrational beliefs one finds in religions these days (flying horses, talking donkeys, magic golden plates, etc), I find the concept of the soul to be less off-putting. And if everyone kept their belief about it to themselves, there'd be no issue. But they don't. They make grave decisions based on their belief. There are certain members of certain religions who, if they die in the process of killing the enemies of their god, believe that their soul will be richly rewarded with scores of virgins (although it's not clear what the souls of female suicide bombers receive).

Since a major theory of this book is that, by believing in magical, irrational things you're missing out on something wonderful, it's a fair question to ask: *What exactly am I missing out on? An eternity of non-existence? The joy of knowing that death is final?* I'll admit, it's not so wonderful sounding. But denying something doesn't make it less true. Besides, there's some value in being honest, isn't there? Isn't it time we put away our fairy tales and take a look around at the very finite landscape of our lives? Because no matter what fantasy you want to console yourself with, this really *is* all you're going to get. So, go live the best life you can.

DISAGREE:
The soul, of course, is just another label for the mind, which *itself* is non-existent. To put it into the kind of simplistic analogies you see on standardized tests:

The MIND is to the BRAIN as WORK is to the OFFICE.

That is to say, the mind is the merely the name for all the activity that goes on in the brain. When the brain ceases to function, the mind (and its spiritual doppelganger "the soul") exists no longer.

Ultimately, the soul is just one of those beliefs that we can label as, "I really *want* it to be true, so it *just has to be!*" But remember the main lesson from the above discussion: There is no unexplained phenomenon for which the postulation of a soul becomes necessary. A slightly depressing reality, perhaps. But I'll take that over a blissful delusion any day.

STATEMENT: Organic food is healthier and safer than conventionally grown food.

AGREE:

I confess, I used to fall for the marketing claims made by the organic foods industry. Every time I made the decision to buy the organic version of some product, it was based on one of the beliefs peddled by that industry:

> *Organic food is healthier for you because it has more nutrients!*
>
> *Organic food is safer because we don't really use pesticides!*
>
> *Organic farming is better for the planet!*

These are three different claims, but before examining any of them, we need to define our terms. If you and I define the word *organic* in significantly different ways, then any discussion we have about organic food is meaningless. Unfortunately, there is no single, agreed upon definition of *organic*. For insight, I turned to the organic industry itself, to see what they consider *organic* to mean:

Simply stated, organic produce and other ingredients are grown without the use of pesticides. (Organic.org)

Really? No pesticides whatsoever? Odd they would say that, inasmuch as *directly below that* on the very same webpage, they give the definition from the United States Department of Agriculture which includes this line:

Organic food is produced without using most conventional pesticides.

I guess the enormous difference in those two definitions escaped the owners of Organic.org. Because of the discrepancy in definitions, let's get another opinion in the search for a consensus. Here's part of the definition of *organic* from the Organic Trade Association (OTA):

Organic food production is based on a system of farming that maintains and replenishes soil fertility without the use of toxic and persistent pesticides and fertilizers. (OTA.com/faq)

Allow me to repeat the key terms side-by-side:

Organic.org tells us: ...*without the use of pesticides.*

The USDA tells us: ...*without using most conventional pesticides.*

The OTA tells us: ...*without the use of toxic and persistent pesticides and fertilizers.*

I would ask the OTA to define *toxic*. Toxic to who? *Everything* is toxic; it all depends on the dose. But in any case, despite what Organic.org would have us believe, organic food is grown using all sorts of pesticides, though not all the same ones used by conventional farmers. Either way, the FDA works very hard to ensure that the residual amount of pesticides in our food—organic and conventional—is well within safety limits.

But, fine. Let's assume that organically bought produce has less pesticide residue compared to conventionally grown produce. The belief I'd like you to reconsider is that, as a result, organic food is somehow *healthier.* As this chapter encourages you to ask: *What reason is there to believe that?* Have there been peer reviewed, scientific studies confirming this claim? As it turns out: *No,* the consensus of experts is that there is no reason to believe organic food is any healthier than conventionally grown food. For example, according to a 2009 study titled <u>Nutritional Quality of Organic Foods: A Systematic Review</u> published in the American Journal of Clinical Nutrition...

> On the basis of a systematic review of studies of satisfactory quality, there is no evidence of a difference in nutrient quality between organically and conventionally produced foodstuffs. (Dangour)

There's also this study, published in the Annals of Internal Medicine: <u>Are Organic Foods Safer or Healthier Than Conventional Alternatives?: A Systematic Review</u>

> The published literature lacks strong evidence that organic foods are significantly more nutritious than conventional foods. (Smith-Spangler)

So, what's the harm in this belief? Or, as Chapter 2 would have phrased it: **What wonderful thing are you missing out on?** Well, mostly you're missing out on money that you could have spent on something else. By all accounts, organic foods cost significantly more than their conventional counterparts. How much extra money are you spending each month for food that has no measurably equivalent benefits? $100? $200? How much is that per year? What else could you have spent that on?

At this point, the devout followers of *organic-ism* (the organic way of life is very much like a religion for some people) will claim that if they switched back to conventional foods they'd be missing out on the wonderfully superior taste of organic food. Sorry, but even that claim is in doubt. No need to take my word for it, though. Have an impartial friend conduct a blind taste test for you using a few samples of the freshest conventionally grown produce, and similarly fresh organic items. If the results are like other taste tests that have been conducted, you won't be able to tell the difference.

As for the claim that organic farming is somehow better for the planet, that's open to debate. Since organic farming is less efficient than conventional methods, they require more land for the same yield. When you have limited area to farm and a world population of seven billion and counting, something has to give. Do we let billions starve in order to minimize the use of certain pesticides? Or do we cut down vast swaths of our ever-dwindling forests, to convert the land to organic farms? I don't have the answers, but at the very least I'd ask you to cast a skeptical eye towards the claims of the organic food industry.

DISAGREE

If money were no object, I suppose I'd have a farm in my backyard where skilled farmers grew all my food without a single added pesticide or chemical. But money *is* an object, a very valuable one, and not something I'm eager to throw away on unsubstantiated claims. And speaking of claims, look at how the Organic Trade Association phrases the purported benefits of organic food:

> There is mounting evidence that organically
> grown fruits, vegetables and grains may
> offer more of some nutrients.
> (OTA.com/nutrition)

And again, on the FAQ page of their website they tell us...

> There is mounting evidence at this time to
> suggest that organically produced foods may
> be more nutritious. (OTA.com/faq)

They sure like the phrase "mounting evidence," but did you notice their use of "may be" in both claims? Let that sink in for a moment and then join me in a primal scream of frustration: *Arrgh!!* Their wording is misleading to the point of being disingenuous! You might as well say, "There is ***incontrovertible evidence*** that organic food *may be* more nutritious." We're still stuck with the qualifying phrase "may be." In fact, you can just as accurately say the following:

> There is mounting evidence that organic
> food may ***not*** be more nutritious.

There is no good evidence either way.

Ask yourself ***how would you go about proving health benefits?*** What test would you conduct? Think of all the contributing factors to our health beyond just what we eat. Even if a convincing correlation is one day shown between an organic diet and long life, that won't be proof that the organic diet was the ***cause***. It might be that people who tend to eat organic food also tend to eat more fruits and vegetables to begin with. They're also much less likely to smoke, and more likely to exercise regularly. That is, had they instead eaten conventionally grown foods their whole lives, it's quite possible they would've lived just as long and healthily due to their overall healthy lifestyle. (And would've had a lot more money for other, more rewarding pursuits.)

STATEMENT: The universe was created last Thursday at 1:00 AM. Our memories were created at that moment, and everything was made to seem as if the universe had been around for a very long time.

AGREE:

If you adhere to the above creed, then you are a member of a religion known as Last Thursdayism. It's not a religion in the traditional sense with churches and ceremonies, but it is a valid philosophical hypothesis. If literal Christians can claim that the world was created 6,000 years ago with the appearance of having actually been made billions of years ago (with light already on its way from distant galaxies, and with a vast fossil record indicating a multi-billion year history of life on Earth), then what stops us from declaring that it was actually made last Thursday?

I asked a Biblical creationist what he thought about Last Thursdayism and he told me that God gives us the free will to believe whatever we want to, but that I was overlooking one very important difference between the two ideas: One of them is the word of God as described in the Bible. (I'll be honest, though: I was expecting him to say, "The world was magically created a few days ago? *That's ridiculous!*")

75

DISAGREE:

Sure, we all disagree with Last Thursdayism, but can you disprove it? Last Thursdayism is similar to the idea that the world is really just a program on an alien super-computer, (a la The Matrix). The point of both of those hypotheses is that our observations may not actually match with reality. Think of the ancients whose observations told them the earth was flat. If you shared with them your absurd hypothesis that the earth was actually *round*, and that no one falls off it because there is no true up or down but instead a gravitational center, you'd be laughed right out of the village. Is Last Thursdayism any different? It, too, seems like an absurd idea, but perhaps we're unable to see the big picture, just like the ancients.

The important difference is this: There's no *reason* to believe the world was created Last Thursday. There's no unexplained phenomenon that such a hypothesis is shedding light on, nor is there any mystery that Last Thursdayism finally solves. The ancients, on the other hand, had good reason to question the long held belief that the earth is flat because certain things didn't make sense. (Why do the masts of tall ships sink below the horizon? And where is that curved shadow on the moon coming from, that you see during lunar eclipses?)

And remember this: Whether we're dealing with someone who declares that the world was created last Thursday, or 6,000 years ago, it's not our job to disprove their ridiculous claim. The onus is on *them* to prove it.

NOTE:
To question the statement below is to question your own Christianity, so please make sure you're up to the task. If you're not ready to do this, I understand. In which case, please skip to the next statement (where we question the belief that the HIV virus does not cause AIDS.)

STATEMENT: The Bible is the unerring word of God.

AGREE:
I'm not going to try to disprove the Bible or look for errors in it. That's been done many times by people far more qualified than I, and yet where did it get us? You still believe the Bible is the perfect word of God. So, clearly, the debunking approach isn't working. The reason is, the people who write those books which show all the inconsistencies in the Bible, and all the factual errors in it, and all the terrible things that the Bible advocates like rape and child murder and slavery and all that— they're not writing those books for believers. They're writing those books for those who *already know* the Bible is the deeply flawed work of ancient, ignorant men.

So, let's forget that approach of attacking the Bible. Instead, as we ask again and again through this chapter, I'd like to know: *What reason is there to believe that the Bible is the unerring word of God?* Which unexplained phenomenon is your claim shedding light on?

* Is the Bible written on **miraculous paper** which hasn't deteriorated in 2,000 years?

* Does the Bible contain **miraculous information**, like how to cure cancer or repair broken spinal cords?

* Does the Bible contain **miraculous warnings of upcoming natural disasters** by giving specific dates and locations? (For example, does the Bible say, "Hey, New York and New Jersey residents: Watch out for Hurricane Sandy coming ashore on October 29th, 2012.")

* Does the Bible **miraculously translate itself** into the language of the person who's reading it? (My wife is Ukrainian, yet when she looks at my English Bible, God doesn't translate it into Ukrainian for her. Doesn't He want her to read it?)

* When you open the Bible, does it **miraculously read itself out loud** to you? This book is supposedly the creation of an all-powerful super-being, so why isn't it able to talk? God made talking snakes, talking donkeys, talking bushes...why not the logical idea of a talking Bible so that blind people can also benefit from His stories about incest and human sacrifice?

When I ask my Christian friends these kinds of questions, they always tell me I'm "missing the point." They then respond with their own vague reasons for believing the Bible was made by an all-powerful being. "I get spiritual enrichment from it," one friend told me. "When I'm reading the Bible I connect with God's covenant," said another Christian friend, obviously assuming I'd understand what on earth he was talking about. But the most troublesome response is when a Christian says, "The Bible is my moral compass." It's troublesome, but it does explain things. If people are getting their morals from the examples set forth in the Bible, then that would explain why we see so much murder and rape and genocide and slavery in the world. People are just doing what the Bible teaches.

I know it's a losing battle trying to convince you that the Bible is, for the most part, a deeply flawed and hateful collection of ancient writings. Still, on the off-chance that you're one of those rare, open-minded Christians who is actually willing to take an impartial look at the gruesome things their beloved book contains, I will list a few resources for you to peruse at the end of this section.

Remember: With every irrational belief you hold, you're missing out on something wonderful. One day, when you finally break free from the Bible's control, you'll feel the relief of being able to think for yourself. You'll discover the joy of learning about the real world. And you'll gain the power that comes from realizing that your value as a person isn't given to you by some book, but comes from your unique personality.

RESOURCES:

BOOKS
Jesus, Interrupted: Revealing the Hidden Contradictions in the Bible (And Why We Don't Know About Them) by Bart D. Ehrman. (Ehrman is a professor of Religious Studies at the University of North Carolina, Chapel Hill, and is a leading authority on the Bible and the life of Jesus. The book is a great read!)

WEBSITES
http://www.thethinkingatheist.com/page/bible-contradictions

http://godisimaginary.com/i5.htm

http://bligbi.com/2007/03/09/10-reasons-not-to-believe-the-bible/

http://atheism.about.com/od/biblecontradictionserror/Bible_Co
ntradictions_Errors_Bible_is_Full_of_Contradictions_Errors.ht
m

DISAGREE:

Karl Marx labeled religion the opiate of the masses, but I view religion as a successfully marketed product. And like all good marketers, those who sell Christianity tell their customers that membership will bring them the greatest pleasure and let them avoid the most horrendous pain. (I explained this idea of maximizing pleasure and minimizing pain back in Chapter 2, but key points are like nails; they need to be hammered in.) When you realize that Christianity is actually a product, then you can see how belief in it provides the customer with multiple sources of pleasure, and prevents multiple forms of pain. Just look at the bullet points...

Your Christian membership provides you with:

* Pleasure from belonging to a large community of people who bought into the same product you did!
Basic membership grants you access to a special building where you're treated like a good customer.

* Pleasure in the promise of an afterlife!
Heaven is the ultimate VIP room, but access is only granted to lifetime members.

* Pleasure from knowing your invisible super-father has a great plan for you!
Membership automatically makes you important and successful, regardless of your current job and status.

* Pleasure from learning that you're one of the chosen ones!
This offer isn't available to just anyone. You need to be able to believe in things without seeing them, and despite all evidence to the contrary. If you can do that, you're a special person and we want you in the club.

Equally powerful is Christianity's message of pain-avoidance.

Your Christian membership also allows you to:

> *** Avoid the pain of eternal suffering in Hell!**
> *Why take the risk of being subjected to constant, excruciating pain without end?*
>
> *** Avoid the pain of being an outcast in your community!**
> *Why suffer the embarrassment of being labeled a non-believer?*
>
> *** Avoid the pain of having to learn how the world actually works!**
> *Physics and chemistry and biology and astronomy all represent **years of study.** With us, there's nothing to learn!*

Of course, by renewing your membership weekly, you can also...

> *** Avoid the pain of admitting you're wrong!**
> *Do you really think you're strong enough to tell all your friends and family that they are wrong? You'd also be admitting that **you've been wrong** this whole time, too! Plus, look how much you've invested in your membership with us. You've got crosses and Bibles and Jesus posters all over your house. You've got one of our logos (the cross or the fish) on your car's bumper. Heck, you're probably wearing our cross-logo around your neck at this very moment. You might even have it **tattooed** on your skin! **You're in too deep!***

Not only do they cleverly market the benefits of membership, Christianity also boasts thousands of superstar endorsers. Is it any wonder their brand is so successful given all their vocal, high-profile members like TV preachers, country singers, athletes, actors and politicians? You can sit down with a Christian friend, open their membership manual (the Bible) and point out all the examples of torture, rape, genocide, baby murder, stonings, beatings, and slavery until your finger is raw from all that pointing, but you will never get him to relinquish his membership. It's like asking a drowning person to give up their life preserver. Without Christianity, they'd be forced to swim in the cold sea of reality.

As always, there's a lot at stake here. Worldwide there are countless thousands of bright minds—Biblical scholars, Torah scholars, Koran scholars, and so on—pouring over these ancient texts, spending literally millions of man hours researching pieces of historical fiction. *These are incredibly smart people* dedicating their lives to tales about talking snakes, flying horses, and magical sky fairies. The world is missing out on all sorts of potentially wonderful things that these people could be doing, had they instead put their minds to work solving real problems.

STATEMENT: AIDS is not caused by HIV but by some other combinations of factors.

AGREE:
People who agree with this have been given the label "AIDS denialist" which is a broad term that covers other dissenting views, such as: *HIV is not an actual virus; HIV is not sexually transmittable; HIV tests are unreliable; standard AIDS treatment is ineffective and even harmful.* Those are all quite different statements, so to be clear, I'm addressing what seems to be the core belief in the denialism movement: "AIDS is not caused by HIV."

The AIDS denialism movement is small but vocal, having spawned numerous websites, books, and at least one documentary (House of Numbers). Of all the incredible beliefs addressed in this book, this is one of the most damaging. You may recall that during his time in office, the former President of South Africa, Thabo Mbeki, maintained an official AIDS denialism stance, and thus his administration did not provide standard treatment for AIDS patients. According to a study by Harvard medical researchers, Mbeki's public health policy resulted in the premature deaths of over 300,000 South Africans. (Dugger)

If we apply the lesson of the previous chapter, it's fair to say that the denialists are missing out on the most wonderful thing of all: *Life!* Someone at risk might ignore safety precautions and contract the virus or pass it on to someone else, or an infected person might rely on unproven, alternative treatments and die prematurely. Such is the devastating power of a single belief.

Because of the non-linear nature of this book, I need to frequently restate that my goal is not to debunk people's incredible claims. (For one source out of many that debunks the claims of AIDS denialists, please visit Aidstruth.org.) Instead, I hope that those with incredible beliefs might take the time to view them skeptically and perhaps reconsider.

If it's your view that HIV is a harmless virus that a person with a healthy immune system would quickly overcome, then *why not volunteer to test your theory by receiving an injection of HIV?* Think of the benefit to mankind if such a test were done on a clinical scale and demonstrated your claim that HIV is harmless. It should be easy to find 1,000 AIDS deniers (the ones who specifically state that HIV is harmless and does not cause AIDS). Inject all of them with varying amounts of HIV and see what happens, if anything, over time.

I'm compelled to immediately follow with the disclaimer: *Please don't!* As in, please don't inject or in any way expose yourself or anyone else to the HIV virus. My suggestion is merely intended to make you think: *How certain are you that HIV is harmless?*

But a more serious question is: *Why not take the responsible people to court?* You have all this "proof" that HIV is harmless, and that standard AIDS treatment is ineffective and even harmful. You have "evidence" that the drug companies are actually responsible for spreading the lies about AIDS in order to sell their expensive drugs, and so on. You have clear cut, indisputable proof, right? (Or else, why on earth would you maintain your belief?) So it should be an incredibly easy case to win. You shouldn't have any problem finding an eager lawyer willing to work pro bono on such a landmark case. And you, too, would reap millions by finally exposing the scandal. So what's stopping you?

If you are an expert in virology or immunology, why not submit a paper to be reviewed by your peers proving that HIV is harmless? (If I'm mistaken, and such a paper exists, I'd love to read it. Please send the name of the article, plus the date and journal of publication to WhatIfYoureWrong@gmail.com). And if you're *not* an expert, then why side with the minority opinion? **What reason is there to believe them?**

The bottom line seems like common sense to me: Whenever we're not qualified to look at the data and make meaningful conclusions, we must rely instead on expert opinion. And on the matter of what causes AIDS, the consensus opinion of virologists and immunologists who have spent years studying the relationship is that HIV is indeed the cause.

DISAGREE:
The AIDS denialism movement is really just a classic conspiracy, and includes the standard conspiratorial elements:

* **Placement of blame on an elite group.** (This can be the government, the media, the Big Pharmaceutical companies, etc.)

* **A very generalized yet much more complicated alternative explanation.** (How exactly did NASA fake the moon landing? How exactly did the Bush administration orchestrate the attacks of 9/11? How exactly did the conspirators convince AIDS researchers worldwide to falsify their findings?)

* A disproportionately high valuation given to the dissenting minority opinion

* The purported "expert testimony" from the minority opinion inevitably has not passed peer review, and is usually the opinion of experts from a non-related field.

That last point is perhaps the most important. *If a dissenting view passes the stringent peer review process, then it becomes the mainstream view*. That's how science works.

Think about ulcers. When I was growing up, the following was the consensus opinion about ulcers:

> <u>STATEMENT</u>: The main causes of stomach ulcers are excess acid in the stomach, stress, and the consumption of spicy foods.

Turns out, the consensus medical opinion was wrong. Ulcers were eventually proven to be caused mostly by bacteria, and the current therapy is now antibiotics. (National Institute of Health) The point being: **There was no conspiracy**. As new evidence came in, the scientific community was happy to change its opinion. The same would happen with AIDS if new research were to ever come in that disproves the current theory.

STATEMENT: The attacks of 9/11 were part of a U.S. government conspiracy.

AGREE:
As you've noticed by now, this isn't a debunking book. There are any number of expert rebuttals that dissect each of the major points in the 9/11 conspiracist's argument. Search the net for "9/11 conspiracy rebuttal" and you'll find plenty to choose from. (My favorite: www.debunking911.com) Remember our goal: I'd like you to reexamine your belief. To that end, just as I did with the moon landing hoax statement in the previous chapter, I'd like to present to you a series of questions:

How did you first come to believe that the events of 9/11 were the direct actions of U.S. government officials?
It's important to see how we come to any particular belief. So tell me, were you one of the demolition experts the government hired to place explosives on each floor of the Twin Towers? Were you on site at the Pentagon shortly after the crash, and that's how you "know" there were no pieces of plane wreckage there?

Do you have any direct observations at all that the attacks were orchestrated by our government?
No, you don't. Just like with the moon hoax conspiracy, all you have are questions. Lots and lots of questions. More on those in a moment, but I want to get to the bottom of how you first came to your 9/11 conspiracy conclusion.

Who planted the seed of doubt?
Did you happen upon a 9/11 conspiracy-themed website, or did a fellow conspiracy-minded friend turn you on to "the truth?"

What difference does it make? Well, if you're honest, you should admit that you have a tendency to rely on false authority. It's not like the website you got your "information" from was any of the following: Popular Mechanics, NSF (National Science Foundation), or the AAAS (American Association for the Advancement of Science). And I'm sorry, but not one of your "experts" has published a peer-reviewed paper proving any aspect of the 9/11 conspiracy.

Why not take the responsible government officials to court?
You have all this "proof" about "squibs" being planted in the Towers. You have proof that the buildings couldn't have collapsed on their own. You have all this amazing, incontrovertible evidence proving the guilt of Bush and Cheney, so why not take them to court? (The wonderful thing about America is you can sue anybody!) Your evidence is rock solid, isn't it? Otherwise, why on earth would you believe it? So it should be an incredibly easy case. Lawyers love big cases, especially the easy ones, so you shouldn't have any problem finding an eager lawyer willing to work pro bono. What's stopping you?

Have you honestly taken the time to carefully research the answers to your 9/11 questions?
The physics of explosions and structural stress and high-impact collisions....these are tricky things, so I understand why you might have trouble comprehending certain aspects of them. I'm certainly no expert, but I carefully read the explanations from

certified science experts. When you ask questions like, *How could the buildings fall if steel only melts at 2,750 degrees Fahrenheit?* or *What were those explosions on each floor just before the tower collapsed?* and so on, what you're really saying is, "I'm not smart enough to understand how steel loses its strength at temperatures as low as 400 degrees." "I'm not smart enough to understand that as the upper floors collapsed, the air pressure blew out the windows on the floor below."

Just as with the moon hoaxers, you guys resemble the religious fundamentalists who use this same **argument from ignorance** (as this fallacy is called) to peddle their creationist views. "I'm not smart enough to understand how evolution works, so I'll just say 'God did it.'" Instead of proclaiming your ignorance, you should be reading actual scientific articles which explain all the things you don't understand.

Why don't you ever ask similar questions of your own theory?
It's good to ask questions like the ones above, but you need to be fair. Why do you not take aim at your own theory? Imagine if your 9/11 conspiracy view somehow becomes the mainstream view. You'd tear it to pieces, wouldn't you?

* Such a conspiracy would involve thousands of people. Why has no one come forward?

* Wouldn't there have been an easier plan involving far fewer deaths that would've achieved the same goals? (Whatever you feel the goals were.)

* Was Bin Laden's first attack on the World Trade Center in February, 1993 also part of the conspiracy?

You don't ask these kinds of questions about your own theory because, despite the nickname for your cause, you're not actually interested in the truth. Instead, you're interested in feeling like you have secret knowledge that few others do.

Ultimately, as we've been seeing with so many incredible claims, there's just no *reason* to believe it. There really is no unexplained phenomena that you're shedding light on. Am I saying there are no conspiracies in the world? Of course not. Conspiracies do sometimes happen, at various levels. But this one takes the cake. The only one that rivals it in terms of complexity is the moon landing conspiracy. But remember Occam's Razor! There is a much, much, *much* simpler explanation for the events of 9/11 than your massive conspiracy.

DISAGREE

Every generation needs a good conspiracy, I guess. While the moon landing hoax is slowly losing steam, the 9/11 conspiracy continues to simmer. Apparently this 9/11 conspiracy theory has a dedicated following, though they spread their suspicions almost exclusively via the internet. I first heard about it from that same friend of mine who also thinks the moon landing was a hoax. My friend really is a normal guy but I honestly can't imagine having such a distorted view of the world.

So why do they cling to their belief, despite all the rational answers to their myriad conspiracy questions, and despite having exactly *zero* evidence to back a single one of their outlandish accusations? **To maximize pleasure and minimize pain.** A 9/11 "truther" must get a lot of pleasure out of considering himself an elite member of those who "know the real truth." That's why they first buy in to the theory. But they nurture it and defend it to avoid the pain of admitting they're wrong. And like most conspiracy theories, this isn't some little mistake they'd have to admit to. This is a *gigantic* one, representing years of investment in an incorrect theory. Imagine admitting that everything you've convinced yourself of and shared with others has actually been wrong. That's an awkward, painful thing to admit. And so they avoid the pain by shrugging off your explanations.

CRITICAL THINKING LESSONS
FROM CHAPTER 3

Here are the takeaways from Chapter 3:

* When you encounter an incredible claim, get into the habit of asking, *What reason is there to believe that?*

* What unexplained phenomenon does your claim shed light on?

* If you're making the positive claim, you need to offer evidence in support of it.

PARTING THOUGHT:
How To Create A Conspiracy

In this book we discuss some common conspiracies because they're a prime example of non-critical thinking. If you tend to fall for conspiracy theories, then a great exercise to develop your critical thinking skills is to try creating your own! And don't worry, creating a conspiracy is easy because:

1) You don't need any evidence. In fact, the absolute lack of evidence makes it an airtight conspiracy because it proves the conspirators were very powerful.

2) The "facts" you quote don't actually need to be true. You just need one "expert" to say they are true.

3) You don't need to lay out any details for how the conspiracy was executed. Vague hints are better, because they allow the conspiracy to take shape in the imagination of the reader.

So below, let me pitch you (pun intended) my own conspiracy theory. (Good luck proving it wrong.)

STATEMENT: The Boston Red Sox did not really win the World Series in 2004. Their entire championship run was actually a hoax perpetrated by Major League Baseball.

PROOF:
How do you explain the fact that *no team has ever come back from being three games down in a series?* And yet, gosh, the Red Sox magically were able to beat the New York Yankees, the best team in baseball, in the American League Championship Series. It's simply not possible. The players were obviously paid off to orchestrate the game.

The Red Sox were not good enough to beat the Yankees. If you recall, they needed the Wild Card spot just to squeak into the playoffs to begin with, whereas the Yankees were the division champions, finishing the season far ahead of the Red Sox. It's obvious they had huge help from the Powers-That-Be in baseball!

Isn't it a huge coincidence that the FOX network (which broadcasted the playoffs) happens to be a for-profit business? And by making the Red Sox win, *they just happened to make millions of dollars in profits?* The playoffs were obviously staged to boost ratings!

How do you explain the fact that *the same thing happened in the 1919 World Series* with—you guessed it—the Black Sox! Gee, now there's a coincidence. (Couldn't the Baseball Execs and the Media Moguls be a little more creative when choosing which teams they want to manipulate?) It's happened before. They're just re-using their own idea.

*   *   *

Creating a conspiracy theory as I did above isn't just an exercise in mockery. By trying to think like a conspiracy advocate, you begin to see all the fallacies they employ to construct their "argument."

CHAPTER 4:
Extraordinary Claims Require Extraordinary Evidence

INTRODUCTION:

In Chapter 4 we'll be discussing mostly pop culture beliefs. We'll look at claims of alien abductions, talking chimps, Big Brother recording all our phone calls, and one of my own bizarre beliefs regarding the universe having once been smaller than a grain of sand. The title of the chapter comes from the oft-quoted line by Carl Sagan: "Extraordinary claims require extraordinary evidence," which is really just a shorthand way of saying, "All scientific claims require the same high standards of evidence." With that in mind, please read the next statement.

STATEMENT: Eyewitness accounts are sufficient evidence to declare something as fact.

AGREE:
Be careful with this one. It's not saying, "Eyewitness accounts are *sometimes* sufficient," and it's not saying they "*can be* sufficient." It's simply saying that they *are*, and that implies *always*. So we need to think about that again. Are eyewitness accounts *always* sufficient evidence to declare something as fact?

Imagine your three-year-old comes rambling into the kitchen, "Mommy, I just saw a dragon and he was flying over our house!" There's your eyewitness account. So we should now believe that flying dragons exist? Or, pretend that a man you don't know tells you an angel appeared to him as he was walking alone in the forest. Would you accept it as fact that he's had direct communication with one of God's servants, or would you be at least a little skeptical of his claim?

I suppose it's a good thing if you originally agreed with the topic statement because it means you're a very trusting person. My guess is that you trust people to tell the truth because you *yourself* tell the truth. But as we think more about it, there are times to distrust the things that even the most trustworthy people claim to have seen. *It has nothing to do with lying.* It has to do with how memories are first formed, how they're filed, and how they're recalled.

Think about a recent trip to the store where you stocked up on groceries. Can you remember the order in which the clerk scanned each item as you were checking out? You can't? Why not? You were there. It was happening right in front of you! *You were an eyewitness.* Okay, never mind the groceries. Can you just describe each of the people you saw while at the grocery store? What was each person wearing? ("Sweater and jeans" won't suffice. I want color, pattern, brand and fabric, too.)

You weren't paying attention at the grocery store? Fine. Then, think back to the dinner you had exactly one year ago to the day. I assume you were paying attention as you ate. Can you please recount everything you ate, in order? For example, first you had two bites of your Chicken Kiev, then a forkful of mashed potatoes, and so on.

Can you recall every phone call you made last month, in order? By "remember a phone call" I mean listing the following information: The phone number you dialed, the person you spoke to, and every word each of you spoke (*verbatim*, not "the gist" of the call).

I'm pointing out the obvious: Human memory is a woefully inaccurate system for recording and storing data. And that's why eyewitness accounts are in *no way* sufficient evidence to declare something as fact. We don't have to look any further than our own legal system for proof of that. Due to faulty eyewitness testimony (among other errors), hundreds of people have been convicted of crimes they did not commit, only to be proven not-guilty years later due to DNA testing. Thousands more likely remain locked up, waiting for their chance at being vindicated.

If you now agree that sometimes eyewitness accounts are *not* sufficient evidence to declare something as fact, then please read the "DISAGREE" section below.

DISAGREE:
Despite everything I said in the "AGREE" section above, this is a tough one for me. If I'm sitting in a windowless basement and my wife comes downstairs saying, "It's raining outside," her eyewitness statement is all the evidence I'd need to believe that it is indeed raining. But even if hundreds of people claim to have seen Bigfoot or the Loch Ness monster, that's not enough evidence for me...*even if my wife were one of the witnesses.*

Why is that? I guess it has to do with the nature of the claim. **Extraordinary claims require extraordinary evidence,** wouldn't you agree? Personally, I'd love to think that there remain a few large, undiscovered creatures still roaming the planet. There certainly are thousands of *tiny* ones that we haven't yet encountered and labeled. But to believe that an undiscovered primate larger than a man lives in the forests of Oregon and Canada? Or that a 30 foot sea creature lives trapped in a Scottish lake? I'm sorry, but I'm going to need to see an actual specimen. Alive, ideally, but even the remains would probably suffice. Don't you think that's reasonable?

If you do agree that such a position is reasonable, doesn't it make you wonder what beliefs you *yourself* hold based only on the eyewitness testimony of others? I know people who believe in alien abductions not because they themselves were abducted, but only because they read the book Communion by Whitley Strieber, or similar stories. Most people who believe in ghosts have never actually had a ghostly encounter themselves. Instead, they accept the eyewitness testimony of others who claim to have seen ghosts. And I know many Christians who believe the accounts described in the Bible because there were "eyewitnesses."

These are all extraordinary claims:

* Bigfoot is real!

* Aliens are abducting people!

* Ghosts exist!

* Jesus rose from the dead!

They can not be accepted as fact on eyewitness testimony alone.

Again, I'm not suggesting you demand evidence from any person who makes any kind of observational statement, like, "It's snowing!" Instead, reserve your skepticism for the bigger claims like, "They're holding alien corpses in Area 51!" Because there's a growing body of evidence which shows that *people make lousy eyewitnesses.* And it's relevant to all of us, if only to the extent that we may one day find ourselves sitting on a jury, having to decide someone's fate based on what someone *else* thinks they saw.

The axiom, "**Extraordinary claims require extraordinary evidence,**" was popularized (and likely coined) by Carl Sagan and has become a catchphrase for skeptics. I use it throughout this book. But it needs to be qualified: *Extraordinary* is a subjective description. What seems like an extraordinary claim to one person might seem reasonable to another. It depends on what you know about the world already. So, think of it as a catchy, shorthand way of saying, "All scientific claims require the same high standards of evidence."

PREFACE:
Pretend you're researching the cause behind a UFO sighting. The person claims to have seen a metallic craft gliding slowly over his neighbor's field one night, and then watched it accelerate away in a flash. He has some grainy photos and shaky video that he took with his cellphone's camera.

STATEMENT: If you *can't* explain what the person claims to have seen, or can't explain the photos or video he took, then that means it *must* be an alien spacecraft.

AGREE:
Imagine that you and a friend create a clever UFO hoax. You concoct a story about seeing a UFO and take some blurry photos of some distant light that I can't conclusively explain. So, since I can not explain what you claim to have seen, and can not explain your blurry photos...that means *you indeed saw an alien spacecraft*? Your hoax has now suddenly become an actual alien invasion, simply because I'm a mediocre investigator working with imperfect information?

There's a Golden Rule in the field of paranormal investigations (which includes everything from aliens to Bigfoot): **My inability to explain your story or photo simply means my investigation skills are limited. It does NOT count in any way as proof of your claim.** All you still have is a blurry photo and a story.

Remember the topic of this chapter: Extraordinary claims require extraordinary evidence. Take a minute or two and ask yourself: **What evidence should we require in order to accept a person's claim that he saw an extraterrestrial craft?**

* Do we accept just his word?

Of course, there are a lot of variables here: Whose word are we supposed to be accepting? Is it the word of your best friend who's normally very clear-headed, or just the word of a stranger on the internet? But assume it's the word of the most trustworthy person you can imagine. Is his word enough?

* Do we accept his word along with photos or videos?

The thing is, how do you know what an *extraterrestrial* craft looks like? How can you be sure you're not looking at an advanced U.S. military prototype, or super-advanced Chinese craft? (This, along with the obvious objection that photos and videos can be faked. And special effects software grows ever more powerful and affordable each year.)

Here's the bottom line: **We can not accept anyone's claim that they saw an alien spacecraft. Ever.** It doesn't matter who it is, or what photo or video "evidence" they have. The reason is: **They haven't proven that the thing they saw is an** *extraterrestrial* **craft**. Actually, they haven't even proven it was any kind of *craft* at all. It could have been some kind of advanced projection, no more real than the images projected on a movie screen. And to me, at least, the term craft implies having space inside for occupants. Have you proven there's space inside the object?

I'm not saying they didn't see *something* spectacular, and I'm willing to accept all sorts of explanations. I'm inclined to believe that the explanation will be simple and mundane, but even if we grant that something incredible is going on, there are still all sorts of possibilities. As I mentioned earlier, the presumed "craft" could actually be an *image* created by an advanced holographic projector. Or, the witnesses could be unwitting victims of a government experiment where they're implanted with fake memories of an "encounter," while equally fake images of the "craft" are planted on their cameras.

Even if we agree they indeed saw some amazing *craft*, they have the separate duty to establish that it's extraterrestrial, because...

- It could be a very advanced Russian or Korean craft.

- It could be a craft created by a privately funded company.

- It could be a craft created by a single, industrious individual.

- It could be a craft from a future human society here on Earth, manned by future humans who are traveling back in time to research their own origins.

- It could be a computer controlled probe sent back in time by a future human society.

- It could be a craft sent from the advanced inhabitants of the underwater civilization of Atlantis.

And so on, and so on. You're making the claim that it was a craft at all, as opposed to, say, a holographic projection. Then you're making the additional unfounded claim it was extraterrestrial in origin. Can't we agree that it's better to just say, "I don't know what I saw," than to make wild speculations?

I'm compelled to clarify that I don't feel it would be *any* of the extraordinary explanations I've listed above, including "extraterrestrial craft." Instead, the unidentified object is inevitably going to be one of the following: Venus, a weather balloon, a lenticular cloud (I saw one of these saucer shaped clouds in Phoenix, once. It was startling!), flares, a meteorite, or any number of other misidentified natural or man-made objects.

DISAGREE:

I, too, disagree with that topic statement. To restate the Golden Rule mentioned above: **My inability to explain your story or photo simply means my investigation skills are limited. It does NOT count in any way as proof of your claim.**

I was talking with a buddy of mine at the gym once about UFO sightings. Up until our discussion he'd been a moderate believer that UFO sightings are sometimes actual sightings of extraterrestrial crafts. (I'm using the standard definition for UFO which refers to *any flying object that a viewer can not identify.*) Presenting the argument laid out in the "AGREE" section above, I got him to concede that no UFO witness has ever proven they'd seen an actual *craft* (because it could've been some kind of projected image), and they'd certainly never proven that the purported craft was *extraterrestrial* (because it could've been from our military, or the Russians, or a private company, etc.) But then my buddy asked a good question:

"So why do so many people think that's what they're seeing? Thousands of people claim to have seen these spaceships. There must be something to it."

Using the terminology of skeptics, this is basically a form of counting the hits and ignoring all the misses. Of all the people who've seen something in the sky that they can't explain, only a small, if vocal, percentage decide they've seen an alien spacecraft. So, my friend's question could just as easily be re-phrased, "Why do so *few* people announce they've seen a spaceship, given all the sightings of unusual aerial phenomena?"

It seems there are a few factors at work, here. First, our brains strive to make sense of the input they receive. If you're walking at night and catch a glimpse of two eyes in the woods that are about chest high, and the faint tip of an antler, your brain fills in the rest of the pattern: **Deer.**

That's a great system for labeling real-world things based on limited sensory information. But the system breaks down when it has no good pattern to match it up with. Not having its own, experienced-based label for the novel and *very limited* input of, say, "large light moving quickly across the sky," the brain seeks any pattern that matches. Of course, it uses the externally generated label from the culture it operates in, hence we call it a, "*UFO.*"

The term "**very limited**" is the key:

* We don't know the size of the object because we don't know how far away it is.

* We don't know how quickly it's moving, either. (Is it small and close, and moving relatively slowly, or bigger and farther away, moving more quickly?)

* In fact, assuming it's an actual object, you probably saw less than 1% of the total possible visual information available had you been perfectly situated to view it under ideal lighting conditions.

Put it another way: If you're standing at one end of a football field and I'm at the other end, and I wave a page from the newspaper for a few seconds, do you think you could recite to me every word that was on that page?

If we had lots of experience with each of the major UFO culprits at close range so their true nature were fully understood just as well as a deer is, then our brains would have accurate labels for those patterns. *Flares*, our brain would say. (Not *UFO*.) But we rarely encounter these phenomena. They're up there and we're down here. And thus our catch-all label is *UFO*.

In this respect, we're like a city-dweller who's come out to the country. He knows almost nothing about animals, but is familiar with monsters. As he's walking one night, he sees the same thing you saw earlier: Two glowing eyes and the tip of a horn. He thinks, *Demon!* and runs away in terror. His brain was working with limited visual information, and had neither the experience nor the label for *Deer*.

But again, most people leave it at that: "I saw a UFO, i.e., an object which seemed to be flying, and which I couldn't identify." They make no further claim as to what it might be, and their interesting story goes no farther than their close circle of friends and family. The trouble is, in our culture the term UFO has become essentially synonymous with "alien spacecraft." And so, seeing the same phenomenon, someone predisposed towards interpreting it as otherworldly will proclaim, "I saw a UFO!" and to them it means an extraterrestrial craft.

STATEMENT: Some people who claim to have been abducted by aliens truly have been.

AGREE:
I used to believe that some purported alien abductions were real. Before reading Whitley Strieber's <u>Communion</u>, which details his claims of having been abducted by non-human entities, I'd heard of Betty and Barney Hill's similar claim. (Back in the 1970's, everyone knew the story of Betty and Barney Hill. Check wikipedia for a synopsis of their story, if you're curious.) But in my twenties I learned to be skeptical. ***A claim of such undeniable importance needs an equal amount of undeniable evidence.***

On the off chance that you're actually one of the people who feels they've been abducted by aliens, I have to ask: *Why haven't you reported it to the police?* Meanwhile, to the rest of you who believe the stories of abductees (but haven't been abducted yourself), you really have to wonder: *Why do abductees or their family members never call the police?* Kidnapping is a federal crime in the United States. By reporting their abduction, these abductees would get the resources of the F.B.I. working to solve it and perhaps prevent other abductions. (Of course, reporting a *fictitious* abduction is also a serious crime.)

You have to admit, it's interesting that when an abduction has an apparent *earthly cause* (that is, the abductors appear to be humans), people call the **police**. Yet when the abduction has an apparent *unearthly cause* (that is, the abductors appear to be non-humans), people call a **psychiatrist**. That's because the police deal with physical evidence and search for external causes. Psychiatrists, on the other hand, deal with memories (real, imagined, or implanted) and search for internal causes.

In any case, it's important to ask yourself: Why do I believe some people's claims about being abducted by aliens? For me, in my pre-skepticism days, it was a numbers game. My reasoning was: *Sure, most stories about alien abduction are either hoaxes or particularly realistic dreams, but some just have to be true.* It's the same reasoning that kept me receptive to the existence of Bigfoot and the Loch Ness Monster. "Some claims just have to be true." On the surface, to a non-critical thinker as I had been, that seemed logical.

It's not.

If we reason that "Some claims just have to be true," we'd have to believe that the Holocaust never happened. Why? Because, sadly, there are a lot of people who claim the Holocaust never happened. Sure, most of them are nuts. But some of them have to be correct, right?

We'd have to believe that the Apollo moon landings never happened for the same reason. Why? Because there are a lot of people who claim the landings were a hoax. Sure, most of those people are nuts. But "reason" tells us that some of them have to be correct.

In fact, the "some of them have to be correct" assertion would need to be applied to all incredible claims with multiple believers. Elvis, therefore, is alive. Aliens built Stonehenge. We all will be reincarnated.

Wrong. *None* of these extraordinary claims have to be correct, no matter how many people make the claim. We should only believe the claims that have evidence to support them.

Another way to reexamine your belief in alien abductions is to play the role of investigator. Story after story comes in to your desk of people being abducted. Obviously, most of them are hoaxes or realistic dreams, but it's your job to investigate all of them. So, where do you start? What exact steps do you take to verify (or falsify) someone's claim? Personally, before starting my investigation, I'd ask:

* Has the abductee or a family member reported this kidnapping to the authorities?
(The answer is always: "No.")

* Were there any witnesses to the abduction other than the supposed abductee?
(The answer is always: "No.")

* Was any evidence found, even *microscopic evidence* of, say, alien skin cells on the person's clothing, or in their room?
(The answer is always: "No.")

* Is there any photographic evidence? (Abductees often claim multiple abductions. One would expect some of them to have security cameras installed in their room.)
(The answer is always: "No.")

Personally, my search for an *external* cause would end there. If I were truly committed to figuring out why the person has made such a claim, I would enlist the help of a trained psychiatrist to determine the internal cause. (For a fascinating read on this subject, I recommend Abducted: How People Come to Believe They Were Kidnapped by Aliens by Susan Clancy.)

DISAGREE:

In all these abduction stories, everyone seems to describe the aliens in the same way: Oval shaped head, big black eyes. *There must be something to that,* says the believer. *They're all seeing the same thing.* To abduction believers, this consistency among the stories lends credence to the general claim. To me, it disproves it. After all, why do they always look exactly alike, which is to say, **remarkably featureless?** No tan skinned aliens? No aliens with wrinkles? No chubby aliens? They never have scars, or acne, or sweat? They are all described suspiciously similar to the alien on the cover of Communion.

There's an ongoing debate, by the way, amongst those who contemplate potential alien forms. Many are skeptical that our unique humanoid form would have evolved elsewhere. I certainly see their point. What are the odds, given all the apparent choices for body types, that these aliens would have two arms, two legs, and fingers? They feel that all these descriptions of humanoid aliens simply belies our innate chauvinism at numerous levels. That is, it's chauvinistic to assume aliens would be carbon based like us, or be DNA based like us, or have heads like we do, and so on.

I used to feel that way too, but I'm wavering. Due to the constructal law, the humanoid form might be one of the few land-based body shapes which allows for the kind of environment manipulation necessary for the development of interplanetary travel. There might be thousands of intelligent alien lifeforms in the universe with all manner of body shapes, but only the ones capable of adroitly manipulating their environment are likely to build spaceships. So, in this one area of the alien abductee story, I won't take issue with the universal description of the aliens as appearing humanoid. The issue to me is that the aliens are always remarkably featureless.

One other thing that I would **not** argue is that it is **impossible** for aliens to have visited here. Yes, I'm aware that even the absolute closest star is roughly 25 trillion miles away. Yes, I realize how much energy it would take to get here, and how long it would take to make the trip, factoring in acceleration and deceleration and so on. Nevertheless, we have to admit the possibility that such a trip could theoretically be done, even if it took 100,000 years, don't we? How about 100 million years? If an alien race is advanced enough to make the trip, I have no problem accepting the fact that they might also have control over how long they live.

And if we're really contemplating all this, who's to say the alien ship had to come from beyond our solar system? If we're being open minded about things, as we should, then we have to admit it's possible there are aliens living on, say, one of the moons within our solar system. In which case, they're not making these trips to Earth from beyond the stars, but merely from their base within our solar system. My point here is: **Skeptics can not go around beating the "It's impossible!" drum**. Alien visitation most certainly *is possible*. As a clear-thinking skeptic you have to concede this, otherwise you're arguing from incredulity: *I'm not smart enough to imagine how aliens could possibly have visited Earth, so therefore it's impossible.* That's not good reasoning.

The reason we should maintain our skepticism is *there's not one molecule of evidence to suggest aliens have ever been here. Period.*

Every now and then you run across a report about the progress being made to teach apes language. Evidently, certain apes like Kanzi the bonobo, Koko the gorilla, and Nim the chimpanzee have been taught to communicate using either American Sign Language or lexigrams (a set of specially designed symbols). Whether it's an article in the N.Y. Times, a report on NPR, a segment on Oprah, or the cover story in TIME magazine, the verdict always seems to be the same: Humans, we're told, are not the sole possessors of language.

It's an incredible claim with profound implications. This whole time we've been assuming that we were the lone inhabitants on this island called language. Seven billion castaways, with no one to talk to but each other. We spend billions with our SETI programs, using gigantic radar ears to listen for a non-human voice among the stars. To hear a non-human *thought*. And it turns out that we needn't have looked any further than the rainforests of Congo.

To think that the only thing holding us back from communicating with another species has been the lack of a patient human trainer who could teach them our language? The news always raised my spirits. But it also raised a big red skeptical flag. My skepticism towards the bold claims that apes are using language has nothing to do with my human chauvinism. I'd love for it to be true. Who wouldn't want to know what an ape is thinking? But wanting something to be true doesn't make it true. So let's examine the following statement...

STATEMENT: Research with certain apes shows that language is not unique to humans.

Sure, an ape can be said to *understand* at least a basic
command-form of language if it correctly carries out novel
requests like, "Give the carrot to the dog," or "Put the ball in
the microwave," but is that all we do with language: Give basic
commands? And is that all we need to declare an ape as
"having language," that it apparently understands such
commands? Has an ape ever been asked a "Why?" question
and responded starting with, "Because..."?

That these apes are able to *communicate* is not in doubt.
Communication is simply the transfer of information, and they
were doing that long before being taken into labs to be taught
hand signals. Monkeys have different warning sounds
depending on which predator is approaching. A gorilla will beat
its chest to warn off ambitious young males. And when
chimpanzees locate food, they will make various grunts and
hoots to let other chimps know where the food source is. So,
yes, certainly apes communicate.

But is communication synonymous with *language?* If so, then
we must declare that ants use language because they lay down
chemical signals which pass specific information to other ants.
And bees have language because the dance they do inside the
hive indicates the location of a food source in relation to the
sun. Heck, molecular biologist Bonnie Bassler discovered that
even bacteria "talk" to each other, using chemical signals to
coordinate their actions and defenses.

Perhaps there *is no* difference between a monkey making the
warning sound for "jaguar" and you and I discussing the origin
of life on Earth. Perhaps language, just like communication, is
simply the transfer of information. To settle the issue, we need
to determine what exactly we mean by "language." As you'll
see throughout this book, I'm keen on getting people to think
for themselves first, before deferring to authority. So take a
moment and think, *How does language differ from mere
communication (if at all)?*

It's a tough thing to nail down, isn't it? You might be thinking, *How am I supposed to know what language is? I'm no expert.* Well, actually you are. We're all language experts. We use language all day, every day, even when we're not saying anything. So give yourself more credit. Take a little time and ask yourself, *What does language do? What do we use it for?* It seems to me that we use language:

...**to maintain social relationships.** (I greet my boss with "Good morning, sir," but I greet my gym buddy with, "Whassup, dude!" Why the different forms?)

...**as a way to elicit information.** ("Do you have the time?")

...**to arrange transactions.** ("If you rake the yard I'll give you ten bucks.")

...**as a fitness display to impress potential mates.** (*This guy is good with words! He must have a great brain and therefore great genes!*)

...**to entertain.** (Think of the works of fiction that fill our libraries. Admittedly, storytelling can also be viewed both as an extreme fitness display, and a business transaction.)

Yet what are the apes using language for in these experiments? On the rare instance that they instigate a "conversation" it's to make what appears to be a direct request: Banana.

I say, "appear" because we don't know the ape's intent. Can you imagine this happening, for example:

1. An ape signs, "Banana."

2. The researcher gives him a delicious ice-cream sandwich instead.

3. The ape puts it down angrily, signing: "I said I wanted a ***banana***, not ice-cream!"

That kind of exchange never happens. When apes communicate, it's for a basic need or want. And when they carry out commands, it's merely to get a reward. The rest, I'm afraid, is the researcher "interpreting" for us. And these researchers are creative interpreters indeed. Take, for example, Bill Fields, a researcher at the Great Ape Trust where the bonobo Kanzi lives. Mr. Fields happens to be missing a finger. "One time," he recalls in an NPR report, "when Kanzi was grooming my hand, when he got to where the missing finger is, he pretended like it was there. And then he used the keyboard, he uttered, 'Hurt?,' as though to say, 'Does it still hurt?'" (Hamilton)

"Hurt" isn't a sentence. It's barely even a statement, and it certainly isn't a question! If I was limited to making just one word "questions" without the benefit of being able to raise my tone at the end of the word, I could still convey it as a question by making a painful expression (I'd bite my lower lip, and wince as I raise my eyebrows in pain), and tapping the spot where the finger is missing: "Hurt?"

So, does their research show that language is not unique to humans? Well, in a major test of Kanzi's ability to understand novel commands (that is, commands it probably hadn't been taught), he got 72% of the 600 commands correct. But the devil is in the details. Take this command, for example:

"Kanzi, take the tomato to the colony room."

Kanzi then took **both the tomato and the orange** to the colony room. His response is labeled "correct" by the researcher because they assume that Kanzi has announced that he wanted to eat the orange along the way. Even if he took only the tomato as he was asked, that doesn't mean he's fully understanding. He hears "tomato", "take" and "colony room." Since you can't take a room anywhere, he's left with having to take the tomato to the room.

Language is my area of expertise, but I don't want to spend

time analyzing their results and interpretations. At the end of this discussion I'll list two articles that examine the issue closely. As always, my goal is only to get you to reexamine your belief that these apes are proof of non-human language.

DISAGREE:
This isn't one of those beliefs that gets challenged every day. I mean, when's the last time you contemplated the notion that an ape is capable of expressing its thoughts through language? But when an incredible claim makes it to the cover of TIME magazine, then it's time to put on the skeptical glasses and take a closer look. As it turns out, to claim that "apes use language" is to strip the word *language* of all the things that elevates it above mere communication.

You can take a look yourself. Just search the internet for "kanzi video" and watch the ones where the person giving commands is wearing a welder's helmet. But if you watch these videos, notice how they are always edited. Why? Because the researchers don't want to show you all the things the ape gets wrong. This is something you'll see again and again from those who are trying to prove an incredible claim: ***They are counting the hits and ignoring** (or in this case, hiding from us) **all the misses.***

Mind you, I'm not putting down the apes. They are clearly capable of associating meaning to hundreds of symbols. But dogs and dolphins and parrots can do that, too. A border collie named Chaser knows the names of over 1,000 items, apparently more than any other animal. The gray parrot Alex also knew hundreds of words and certain commands. Why, then, are apes credited with "language" when a command is phrased as "Get the ball!" but dogs are not given equal credit when we phrase the same request as "Fetch the ball!"

Like so many believers, the ape researchers are passionate and vocal, announcing their claims without first asking the necessary skeptical questions. And I understand that. They love these animals. Most apes are endangered, and they're doing what they can to call attention to their plight. But I would ask these researchers to phrase their claims more modestly. When Alex the parrot was alive, his trainer, Dr. Irene Pepperberg didn't claim that Alex was using "language". She instead more accurately called it "a two-way communications code." (Wise, 90)

Although these long-term experiments to determine whether apes can acquire human language are interesting, I think the real science lies in understanding the actual communication that chimps and monkeys (and whales and dolphins, etc) use in the wild, and in developing theories about the evolution of communication in general.

RESOURCES

For a brief critique regarding the claims of ape researchers, please read Clive Wynne's article *Aping Language: A skeptical analysis of the evidence for nonhuman primate language.*
<http://www.skeptic.com/eskeptic/07-10-31/>

For an in-depth analysis, please read the 40 page abstract *Constituency and Bonobo Comprehension* by Robert Truswell, at the Center for Cognitive Studies, Tufts University.
<http://www.lel.ed.ac.uk/~rtruswel/kanzi3.pdf>

STATEMENT: The U.S. government records and analyzes all telephone conversations made in the United States (including cellphone calls).

AGREE:

I've been hearing this more and more from smart people I trust. But when I corner them, "How do you *know* the government is recording and storing every single phone call?" the answers always sink to the level of what I call **urban legend proof**: *My friend's sister-in-law plays racquetball with a guy who knows one of the engineers who does data storage for a company that works with the NSA.* As I said, my friends are smart. They work at high-tech companies in places like Cambridge, Mass and the Beltway. I don't want to call their beliefs into question, so I nod my head, "I see," and then change the subject, lest I be called "naive."

But in this book I get to vent: "Your friend's sister-in-law just might play racquetball with a guy whose company does some data storage for the NSA. *But how does that prove they're recording every single phone call made by every single American???*"

That is an incredible claim. Aside from the staggering logistics of such a task, if true, it would mean our government is in violation of all sorts of privacy laws. Is it impossible? I suppose not. They've certainly violated privacy laws before, though not anywhere on the scope that's being claimed here. The question we need to ask is, "Is there *reason* to believe that?" Since you agree with the claim, take a moment and ask yourself why. *What evidence have you encountered to convince you?*

Having researched this a bit, it seems there is *some* evidence. Just enough evidence, anyway, to fuel the Orwellian conspiracy theorists:

* During his tenure, President Bush authorized the NSA to conduct warrantless surveillance. They monitored any phone calls, internet activity, text messaging, and so on involving people of interest outside the U.S., *even if the other party was located within the U.S.*

That is indeed troubling, breaking the cardinal rule of NSA surveillance which had been: "Never spy on U.S. citizens on their homefront." It certainly caused an uproar. But it hardly implies the recording of every call.

* In 2006, AT&T technician Mark Klein alleged that AT&T cooperated in an illegal National Security Agency domestic-surveillance program. It led to a lawsuit which was later dismissed, but the intercept room he discovered at the AT&T building in San Francisco, along with the apparent data capturing equipment it contains, does seem to exist. Klein's claim is further backed by NSA whistle-blower William Binney, a former NSA crypto-mathematician. To quote James Bamford's WIRED magazine article from March, 2012, which served as Mr. Binney's soapbox:

> The network of intercept stations goes far
> beyond the single room in an AT&T building
> in San Francisco exposed by a whistle-

blower in 2006. "I think there's 10 to 20 of them," Binney says. "That's not just San Francisco; they have them in the middle of the country and also on the East Coast." (Bamford)

The existence of that "NSA Access Only" room in the AT&T building (room #641A, to be exact) is disturbing. But Binney's elaboration in the WIRED article begs the question: *Are there* **ten** *or are there* **twenty**? That's a huge difference. If you have evidence of any others—be it ten or twenty—why not give the evidence instead of guessing wildly? And most relevant, by Klein's own admission, the NSA seemed to be capturing only internet data, and **not** phone calls. (Klein)

* Mr. Bamford's article mentioned above is titled *The NSA Is Building the Country's Biggest Spy Center (Watch What You Say)* and it describes the construction of something known as the Utah Data Center. It is **alleged** to capture:

> "...all forms of communication, including the complete contents of private emails, cell phone calls, and Internet searches, as well as all sorts of personal data trails—parking receipts, travel itineraries, bookstore purchases, and other digital 'pocket litter'," though its precise purpose is secret. (Bamford)

We're told in the article that it's being built under immense secrecy, which begs another question: *If it's all so incredibly secret, then how did you get your information? How did you come to your conclusions?*

The article also accuses NSA deputy director Chris Inglis, who ran its worldwide day-to-day operations, of engaging in "double-talk" when he stated: "It's a state-of-the-art facility designed to support the intelligence community in its mission to, in turn, enable and protect the nation's cybersecurity."

How is that "double-talk?" Such a term borders on libel. The definition of "double-talk" is: *deliberately unintelligible gibberish*. Would you characterize Mr. Inglis's answer as "unintelligible gibberish?"

In a response to the article written in WIRED magazine, the NSA's spokeswoman Vanee Vines wrote a letter to FOX news:

> "Many allegations have been made about the planned activities of the Utah Data Center. What it *will* be is a state-of-the-art facility designed to support the Intelligence Community's efforts to further strengthen and protect the nation. NSA is the executive agent for the Office of the Director of National Intelligence, and will be the lead agency at the center." (Prann)

Meanwhile, in his WIRED article Mr. Bamford tells us that Binney, the NSA whistle-blower, pinched his thumb and forefinger together: "We are *that far* from a turnkey totalitarian state." But that's another huge accusation. *A totalitarian state?* As in, a government that subordinates the individual to the state and strictly controls all aspects of life by *coercive* measures?

One wonders if Mr. Binney truly understands what he's implying. And that's the problem here. When things are too complicated for us to come to conclusions on our own, we must rely on the opinion of experts. Yet how much stock can we put into what experts like Mr. Binney and Mr. Bamford say when they resort to terms like "double-talk" and "totalitarian state?"

We're really no closer to having the evidence we need to believe that the NSA (or any other branch of the government) is recording all our phone calls. But there *is* this related bit of evidence:

* In 2006, the USA Today reported that the NSA had been collecting phone call *records* of millions of Americans. The article goes on to say:

> This program does not involve the NSA listening to or recording conversations. But the spy agency is using the data to analyze calling patterns in an effort to detect terrorist activity, sources said in separate interviews. (Cauley)

The key phrase there is "phone call records." What that means is they are (evidently) not storing the actual *audio* of each call, but all the related information: Who the caller and receiver are, the date and time of the call, the length of call, and so on. However intrusive that might be, it's not the same thing as recording and storing the audio of every single phone call.

DISAGREE:
One thing that always gets my skeptical juices flowing is when there are variations on a particular incredible claim. It's a sure sign that the people making the claim have no concrete evidence. The classic example is, of course, the claim that there's an all-powerful being ruling the universe: *"We believe there's an all-powerful God named Allah." "No, His name is Jesus!" "No, the name of the all-powerful one is Yahweh." "No, it's Xenu!"* As regards this claim about our government eavesdropping on every single phone call in the United States, some believers declare that the government isn't actually *recording* all our calls. Instead, the claim is that *they're monitoring every phone call in real time*, listening for keywords.

But how could that system work without recording? What if the final word of the conversation is one of the keywords they're on the lookout for? If you didn't record the whole conversation, then what's the benefit? There'd be nothing for the human listener to analyze. And any such monitoring of calls would indeed need a human at some point. Unless the NSA has some ultra-advanced A.I. software which is orders of magnitude better at understanding human speech than what currently exists, then a human needs to listen to all the flagged conversations. Not that I want to argue from ignorance. The NSA has a massive budget, and they likely have the very sharpest minds at their disposal. Despite the enormity of the task, which includes not just intercepting but having to store (at least temporarily) the 4 billion phone calls or so that we make *every day*, it still is possible. But ultimately, we want to ask: *Which unexplained phenomenon is your hypothesis shedding light on? What mystery are you finally solving with your proposition?*

Are millions of Americans being rounded up Orwellian style?

Are our jails overflowing with American dissidents?

Are they even overflowing with terrorist suspects?

Certainly we're dealing with some serious issues here: *National security versus the right to privacy. The good of the people versus the good of the individual.* And some tough questions need to be answered: *What are we willing to sacrifice to lessen the likelihood of future attacks?* And of course: *Who watches the watchers?* But in the absence of any concrete evidence that they're eavesdropping on every single phone call, I'm willing to give my government the benefit of the doubt.

STATEMENT: There is life on other planets.

AGREE:

There is a disturbingly large number of people who "know" that there is life on other planets. Disturbing because the meteorite ALH 84001 notwithstanding (see Chapter 6), there is absolutely no evidence to support this paradigm-shifting belief. Along with the highly respected Dr. David McKay, who is the Chief Scientist for Astrobiology at the Johnson Space Center, you have millions of laypeople making such a claim: Scientologists with their belief in a galactic confederacy; Mormons and their inhabited world of Kolob; not to mention the millions of UFO enthusiasts, alien abductionists, and the proponents of "aliens built the pyramids," and so on.

Of course, we've all contemplated it. Who hasn't looked up at the stars and wondered, "Are we alone?" It's such an appealing notion, we end up lowering our shield of skepticism and embracing the idea. Our logic? *The universe is so big there just* ***has to be*** *other life out there.*

That's not sound reasoning. We have absolutely no idea how likely it is that non-living molecules will come together in such a way as to form a self-replicating molecule. Perhaps it's a one in a trillion in a trillion kind of event. In which case, we may very well be alone. Some wishful astronomers toss around big numbers as they articulate arguments for why life is most likely abundant throughout the universe. Their calculations, of course, include lots of assumptions and speculations. But even if every star in the universe were known to have one Earth-like planet in orbit, it still would leave us no closer to believing there's life out there. ***We still have no idea what the likelihood is that a self-replicating molecule will come to be.*** So, dream all you want. I certainly do. But don't make claims about things that are absolutely unknown.

DISAGREE:
The correct stance here is to neither agree nor disagree, but simply say, "I don't know." (A tough thing for people to do, apparently.) This wasn't always my stance. When I was younger and less skeptical, I used to argue that the universe was most likely teeming with life. I now tell people, "I don't know if there's life elsewhere in the universe. We certainly have found no evidence of it, and we may very well be alone."

I was hoping we'd have a chance to perhaps discover life on Mars with NASA's Curiosity rover (though that wasn't its explicit mission), except some engineers at NASA goofed and didn't re-sterilize the robotic arm when they opened it just prior to launch. Now, the rover is being specifically directed *away* from any signs of water so as not to contaminate the planet with whatever earthly microbes are still on the arm. Good job, guys. (I'm a big NASA fan, but that's a monumental mistake.)

If we *are* alone, it makes each of our lives remarkably, spectacularly special. You are a sentient being which, despite cosmically long odds, has come to exist. A part of the universe (you) is now so highly organized as to be able to ponder its own existence. We may take that for granted, but it boggles the mind when you think about it.

PREFACE:

I do not mean to single out the incredible claims of Christianity, which is why the incredible claims of other religions are also called into question throughout the book. Any extra attention Christianity receives is simply a reflection of having been raised in a predominantly Judeo-Christian nation, and an acknowledgment of my primary audience.

STATEMENT: Jesus died (as a result of his crucifixion) and came back to life.

AGREE:

This is it. The fundamental tenet of Christianity. If you wish to call yourself a Christian, you must accept the claim that Jesus was medically dead and then came back to life. To question this statement is to question your own Christianity, so please make sure you're up to the task. If you're not ready to do this, please skip to the next section (where I question my *own* outrageous belief that everything in the entire universe once occupied an area smaller than a grain of sand.)

Still here? Great! That means you're willing to reconsider this fundamental belief of yours. I doubt many Christians have ever taken a critical look at their belief in Jesus' purported resurrection. For most Christians, belief in the resurrection is like some mysterious object permanently locked away in the bottom drawer of their desk, with a little sign saying: '*Not Subject To Examination!*' But isn't it time we unlock that drawer and shine a flashlight in there? Remember, I'm not asking you to change your belief. We just want to take a look.

124

To start with, for such an incredible claim, we would expect an equally incredible amount of evidence, right? For example, I can't just tell you that my friend's cousin can levitate cars with his mind, while offering you only the testimony of my friend and his cousin. You'd want more than that, and rightly so.

So, do you know what the Christian apologists offer as evidence for Jesus' claimed resurrection? ("Apologists" are Christians who attempt to give proof for their faith beyond simply declaring that "It's written in the Bible.")

* His tomb was discovered to be empty.

* There were over 500 witnesses who saw him alive after he died.

* Many people's lives were changed. (James became a leader; Paul was converted; people even died for their belief.)

I suppose you could add the "proof" that it is written in the Bible. If someone believes every word in the Bible to be true, then that's all the proof they'd need. But be aware that few Christian fundamentalists rely on that argument. For them, the three main points listed above represent their proof: *Empty tomb; 500 witnesses; Lives were changed.* What do you think? Is that iron-clad proof?

If you're a Christian who's trying to rationally analyze their faith for the first time, this can be very awkward. I remember when I first reexamined this belief. I felt like I was doing something wrong. My mind protested: *I shouldn't be asking such questions!* Still, the best thing to do is to let someone ponder it themselves. Here's how to reexamine this belief without feeling like you're doing something wrong:

Pretend you're writing a novel and one of your characters is skeptical of the resurrection story. This way, it's not really you who's asking questions, it's the character. What questions might this skeptical character ask?

125

This is a very important part of the process, so please take a moment and jot down some skeptical, "cross-examination" kinds of questions regarding the Christian offer of proof for the resurrection. Again, I realize this is difficult, but be fair. Try your hardest to find weaknesses in their argument. Nothing less than your life is at stake here.

When you've got some good questions, come back and let's compare notes. (I'll see you in the 'DISAGREE' section below)

DISAGREE:
In this section, we'll continue the exercise described at the end of the 'AGREE' section above; namely:

What questions should a skeptic ask regarding the story of Jesus' resurrection?

1. Isn't it possible (however remote) for Jesus to have survived his torture and subsequent crucifixion?

Aren't human beings capable of surviving incredible things? Think of earthquake victims pulled from the ruble days later. I think, if you're being fair, you would have to admit that it *is* possible for a human to survive crucifixion. After all, Josephus (an independent historian from the 1st century, and whose accounts are generally accepted as accurate) recounts having met *three* people who survived crucifixion.

But even if you maintain that it is impossible for any man to survive crucifixion, *are you saying Jesus was just any man?* Aren't we supposed to consider him an all-powerful god who can walk on water and resurrect dead people? He has the power to come back to life, but not the power to simply survive in the first place?

Yes, I know, such a scenario contradicts the story, but I'm asking: *Is it possible for an all-powerful, supernatural being to survive crucifixion?*

If you answer NO, then your God isn't all-powerful, so it seems to me you have to acquiesce and admit, YES, Jesus could have survived.

2. Regarding the purported empty tomb:

Aren't there other possible explanations for why Jesus' tomb was found empty?

* The distraught women who thought they were at his tomb were actually at the *wrong* tomb.

* Maybe some guys moved the stone and took Jesus' body.

* Maybe Jesus didn't actually die but merely lost consciousness. He then moved the stone away on his own when he regained consciousness. (And/or someone helped him move it.)

Of course, there's one other obvious explanation:

* *Maybe the story isn't true.* Maybe Jesus actually died, and stories about his empty tomb were made up much later.

Even if you insist that the explanation for Jesus' empty tomb has to be supernatural, *who says your supernatural explanation is correct?*

* Maybe a more powerful *second god* took Jesus' dead body from the tomb. This more powerful second god could then have created someone to look like Jesus, to walk around and convince the 500 spectators. (This, by the way, is along the lines of what most Muslims believe.)

In fact, *there's an infinite amount of equally valid supernatural explanations!* Just a few are...

- Thor took the body.

- Egyptian sun god Ra took the body.

- Xenu, the dictator of the "Galactic Confederacy" (of Scientology fame) took the body.

- _____ (insert god name) took the body.

All the above are equally valid supernatural explanations for the empty tomb. A Christian would likely say, "Don't be crazy! Those don't fit the story!" to which I'd respond: *What came first, the facts or the story?*

Another possibility is that an advanced alien race was involved. They beamed down into Jesus' tomb, took a sample of his DNA, beamed back to their ship and quickly made a clone. Or perhaps they were able to repair the damage to his body and restore life through their advanced medical technology. However farfetched you might find this alien hypothesis to be, it is still *less incredible* (because it has a natural cause), than invoking the involvement of a supernatural being.

When we think about it, to go from, "Gosh, the tomb was empty!" and jump to the conclusion, "An all-powerful, supernatural being did it!" is a bit eager, don't you think?

3. Regarding the purported 500 witnesses, shouldn't we ask the following investigative questions:

* How do we know there were over 500? Who did the counting?

* Are we sure no one was counted twice?

* If someone indeed counted the witnesses, then why not tell us the exact number?

* What were their names?

* How credible are these witnesses?

* Did anyone get written testimony from the witnesses, and has it been preserved?

* Why did Jesus only appear to his followers? Why didn't he show himself to his critics like the Jewish leaders, or the Romans...or Pontious Pilate himself?

The above question makes the assumption that Jesus actually rose from the dead and showed himself to anyone. A better way to ask the question would be:

* Isn't it a coincidence that only Jesus' followers were the ones who claimed to have seen him, and that his Jewish critics, or Pontious Pilate never mention it?

* **Isn't it possible that the story is just a metaphor?** Perhaps he didn't literally come back to life, but instead his "spirit," that is, *the good things Jesus stood for* lived on past his death. Doesn't that seem more reasonable?

4. Regarding the "evidence" that many people's lives were changed...

Is that really evidence? Christians love to ask, "Who would die for a lie?" Well, people choose martyrdom all the time. Think of the monks who set themselves on fire. Think of Jonestown. Think of Koresh. People die for lies all the time. If "dying for a lie" is proof, then we're forced to believe that David Koresh was God's final prophet, and that the members of the Heaven's Gate religion are now safely aboard the alien spacecraft they believed was following the Hale-Bopp comet.

129

The thesis of this chapter is that extraordinary claims require extraordinary evidence. As for Jesus' supposed resurrection, there simply is no evidence even remotely commensurate with the claim.

STATEMENT: The entire universe was once smaller than a grain of sand.

AGREE:
If ever there was an extraordinary claim, this is it. It states that all the matter in the universe—all those hundreds of billions of galaxies with their hundreds of billions of stars—was scrunched into an infinitesimally small point. And not just matter, but space *itself* was equally small, way back when. It's an exceedingly bizarre thing to believe, isn't it? So, why do you agree with it? Remember, it's not good enough to say, "Everyone knows the Big Bang is a fact." Even *Einstein* assumed the universe was a constant, unchanging size before Edwin Hubble's radical discovery in the 1920's. (More on that in a bit.)

For a good mental exercise, imagine going back in time to 1910 or so, when Einstein was about 30 yrs old. There he is, naively believing that the universe is a constant size. You catch up to him on the campus of the University of Zurich, "Excuse me, Mr. Einstein? You might not believe this but apparently the entire universe was once smaller than a grain of sand."

"Is that so?" he asks. "What an extraordinary thing to claim. What reason is there to believe that?"

What do you tell him? If you're at a loss, try to think of what mystery the claim sheds light on. "Well, the thing is, sir, it seems that no matter which direction we look out into space, the objects we see are moving away from us with a velocity proportional to their distance away from us. So, the farther away something is, the faster away it's moving."

"I haven't read about that," Einstein says, with a curious tone but a doubtful expression.

"It's not generally known yet," you explain. "But keep your eyes open for the observations of Edwin Hubble over at Mt. Wilson, California."

"You're telling me there's evidence that the universe is expanding?" he asks.

You nod your head. "Apparently so."

Einstein lowers his gaze. He's in thinking-mode. "I understand the implication. If it's expanding going forward in time, then by extrapolation it must get smaller and smaller as you go back in time. But still..." He looks up at you. "*Smaller than a grain of sand?* That would have made it extraordinarily hot back then."

"Exactly!" At last, a chance to use what you learned in your high school science classes. "They've observed the remnants of that early radiation."

"Really? Is it uniform?" Einstein asks.

"I think so?" you answer, hoping you got that one right.

"Of course, it's been cooling down since then." Einstein scratches his head. "Wait....*When was it?*"

"The discovery?" you ask.

"No. The **start**. When was the start of the universe?"

"Oh, about 13 billion years ago-*ish*. Give or take a billion."

"Thirteen billion? *Gott im Himmel!"* He looks up at the sky, getting his mind around it. He looks shaken. "Where can I read about this discovery of the background radiation?"

Knowing that the discovery doesn't come until 1964, nearly a decade after Einstein's death, you gloss over things a bit. "Well, you know how those observational scientists can be. They're still ironing out a few details. But you're right about it cooling down. I think they measured the radiation at three degrees Kelvin."

"How do you know all this? Are you with the Americans?" Einstein asks. "Do they have telescopes we haven't heard about?"

"Yes," you tell him. "And it turns out that these telescopes have supplied still more evidence to support the claim. You've heard of those cloudy, spiral-shaped thingies found throughout the night sky?"

"You mean the spiral nebulae," Einstein clarifies.

"Right. Well, it turns out those things aren't actually gas clouds located **within** our Milky Way galaxy."

"Are you saying..." his voice trails off.

"Yes. Those are actually huge groups of stars—they're other **galaxies**—located far beyond ours."

"And those are the objects you mentioned, that are racing away from us?" This is arguably the brightest man who has ever lived, and yet his eyes shine with a childlike sense of wonder.

133

His understanding of the universe just underwent a tremendous shift.

"Yes. And the farthest galaxies look very different, as measured by their star and quasar formation, than the nearby ones."

"So the universe is clearly *not* in a steady state."

"Exactly."

"Do they know the size?"

"I'm sorry, Mr. Einstein, but I have to be going. There's a book I have to get back to...."

* * *

Those are the three main reasons why I believe that the universe was once incredibly small:

- *The universe is expanding.*

- *No matter where we look we can still see the afterglow from its very hot beginnings.*

- *Things that are farther away look different than things that are closer.*

There is other evidence, too, but those are the three things I can most easily grasp. And of course, it's the current consensus opinion of virtually all astrophysicists.

DISAGREE:

No matter how strange it might be to think that everything we see around us was once contained in such an incredibly small area, the virtually unanimous consensus of experts worldwide is that there's no better explanation for the observations that have been made. I suppose you could choose to believe that some kind of magical, supernatural being created the universe "in medias res" (which is a storytelling term for starting a story "in the middle of the action"), but that is orders of magnitude *more* bizarre, and requires much more explanation. (How did this being do that? Why did it choose to fool us so convincingly? and so on.)

Notice how I avoided using the phrase "Big Bang" in the statement itself. I did so because it's a loaded term that fundamentalists automatically dismiss. I didn't want that. I wanted them to think about the central tenet of the theory. The reason fundamentalists seem to dislike the Big Bang theory is that they mistakenly believe it accounts for the *creation* of the universe. It does not. All the Big Bang theory seeks to explain is *how the universe developed after its creation*. Scientists have no idea why it happened. (They have some guesses, but that's all they are at the moment.)

I also avoided the term "Big Bang" because it's a misleading name. It gives the impression that space already existed, and this densely packed dot simply exploded, filling the space with matter. But that's not how it happened. The theory says that *space itself was microscopic in size*, and then—for reasons unknown—underwent massive inflation incredibly quickly. Sure, the title "Big Bang" is catchy, but we should at least give it the *subtitle*: The "Rapid Expansion From A Singularity" theory. If you truly disagree with the Big Bang theory, there's probably little I can do to change your mind. I would ask that you read the 'AGREE' section, though, just to see the reasons why most people accept it.

It's important to note that rational people are always willing to change their beliefs when new evidence suggests that they do so. In the first half of 20th century, the consensus belief was that the universe was static and unchanging in size. Even Einstein believed it for much of his life. But as new observations came in, they began to realize that the universe was actually expanding. And so that new idea of an expanding universe became the standard belief. That is, until about the year 2000 when new observations showed that not only is the universe expanding, but the expansion is actually ***accelerating***. How counter-intuitive is that? And yet that is now the current belief.

In science, theories are never scrawled in permanent ink. Instead they're written in chalk, to be erased or revised as new evidence is collected. As for the state of the universe, new data may well come in and the idea will get refined once again, or even be changed radically. You need to keep an open mind. Only an irrational person would cling so stubbornly to an ancient belief, unwilling to consider a different viewpoint, unable to think for himself.

STATEMENT: You can't prove anything with 100% certainty.

DISAGREE:

This is surely the believer's final defense against reason: "Everything comes down to faith at some point because *you can't prove anything with 100% certainty!*" insists the believer.

I disagree. If we're talking about mathematical or logical proofs, then you can indeed prove an abstract statement with 100% certainty. For example:

All numbers greater than 2 are also greater than 1.

The number 7 is greater than 2.

Therefore, 7 is greater than 1.

If each of the conditional statements is true, then the conclusion is 100% certain. It has been proven. On the other hand, if we're talking about proving claims about the real world, then I'm more inclined to agree that nothing can be proven with 100% certainty. For more on that, please read the 'AGREE' discussion below.

AGREE:
If we're trying to prove claims about the real world, then nothing can be proven with 100% certainty. Here are two reasons why:

Reason #1: However unlikely it might seem, it's nevertheless possible that what we consider "reality" is actually just a program on an extraordinarily advanced super-computer.

Think that idea is too farfetched to even consider? Look at how much our computers have in advanced in the paltry 75 years since their invention. Can you imagine how advanced they'll be a thousand years from now? How about a million years from now? Compare the earliest, blocky computer games like Pong or PacMan to the brutally realistic games kids play now on their advanced systems, and then multiply that progress by a hundred thousand more generations. All "life" on "Earth" might actually just be some life simulation program on an alien super-computer. Or, we could be in a program created by humans a few thousand years in the future, as a way for them to understand their past. You can not argue from ignorance on this. That is, you can not say, "I don't understand how an extraordinarily advanced civilization could create such a complex computer program, therefore it must not be possible." That's not sound reasoning. We have to accept that it *is* possible and live with the uncertainty.

Mind you, I am *not* making the claim that our reality is merely a computer simulation. There's no reason to believe that, because it does not shed light on any particular unexplained phenomenon. The reason I mention it is that, *given that our reality could be simulated,* then anything we "know about the world" is only about things we know *inside* the program. Outside in the real world of those who created the program, things may be very different. Earth might be fictional. The speed of light might be ten times faster than it is in our simulation. In fact, all the fundamental forces in physics might actually have different values, and so on.

Reason #2: You could be insane, and your "reality" is merely an hallucination. For example, it's possible this book that you think you're "reading" isn't real; you're merely inventing it. In fact, it's possible that all the other things you "see" around you right now are imaginary, too. However distressing it might be to consider, it is nonetheless possible that you are actually lying strapped to a table somewhere in the depths of some mental hospital, your whole world an elaborate and continuous hallucination. *Given that your reality could be a hallucination,* then anything you "know about the world" is only about things *inside* your mind. The real world of the doctors and nurses who are attending to you, and the real world beyond the mental hospital you're currently in, could be very different.

Therefore, no claims about the real world can be proven with 100% certainty, because...

...it's possible our "reality" is being simulated within the computer of an advanced civilization.

...it's possible you are insane and your "reality" is merely an hallucination.

Having broached them for the sake of being literal and thorough, let's now put aside such fantastic hypotheticals like alien computers and elaborate hallucinations. **Assuming that our shared reality is not a simulation, and one's personal reality is not an hallucination**, can we then prove any claim about the real world with 100% certainty?

Some claims, yes. If we can agree on all the terms, then we can set up statements about the real world to function like abstract, mathematical proofs. Claims like, "There is a book on this table." We just need to agree on the terms, *What is a book?, What is a table?* and so on, and we'll be able to prove it with 100% certainty. Barring that, however, most claims about the real world can only be proven with varying degrees of certainty.

But this whole debate makes me want to yell, "So what?!" It hasn't been ***proven*** that, "Every time you drop an object which is heavier than the air it displaces, it will fall to the ground," but do any of us therefore disregard the general premise and step out of our 3rd story window? There is a preponderance of evidence to support our belief in the theory of gravity, the theory of evolution, and the theory of the rapid expansion of the universe. We don't need 100% certainty. What we ***do*** need is sufficient evidence.

So then, what constitutes sufficient proof of a claim? It depends on the field of study which governs related claims. In the field of medicine it's the results of multiple, randomized, double-blind control trials that constitute sufficient proof that something is true, or that some therapy works. That is the extraordinary evidence the consensus of medical experts require for any particular claim.

Meanwhile, in the field of particle physics, sufficient proof of a discovery comes when the probability of the results reaches the 5 sigma level (meaning there is less than 1 chance in 3 million that the observed signals were the result of statistical fluctuations.) This is the extraordinary evidence that experts in quantum mechanics need to believe the claim that, for example, some new fundamental particle has been discovered.

In the field of history, the existence of an historical figure or event is considered sufficiently proven when there's a convergence of massive amounts of evidence from multiple sources. This is true whether we're discussing an historical figure like Julius Caesar, or an event like the Holocaust.

This isn't meant to be a treatise on the forms of sufficient proof in all disciplines. Instead, by giving examples of the kinds of stringent proof required by a variety of experts, ***the goal is to encourage you to insist on equally high standards of proof for any claim that seems incredible to you***. If the theory of evolution seems like an incredible claim to you, then by all means, please ask for the converging evidence that supports it.

(And brace yourself for a deluge!) Likewise, please don't take offense when others ask for equally stringent proof for the claims which you subscribe to.

CRITICAL THINKING LESSONS
FROM CHAPTER 4

Here are the takeaways from Chapter 4:

* My inability to explain your story or photo simply means my investigation skills are limited. It does not count in any way as proof of your claim.

* Be wary of bad science, wishful thinking, and a general lack of skepticism.

* All scientific claims require the same high standards of evidence.

* Insist on high standards of proof for any claim that seems incredible to you.

* Don't take offense when others ask for equally stringent proof for the claims which you subscribe to.

* Historical events are accepted as factual based on a convergence of evidence from multiple sources.

* People make lousy eyewitnesses.

PARTING THOUGHT:
How Do We Convince a Dissenter?

Imagine trying to prove even the simplest of real world claims to someone who hasn't fully developed their critical thinking skills. For example, you're standing outside in a downpour with one-hundred other people. You make the claim, "It's raining," and then ask, "Does everyone agree?" Ninety-nine people concur, but one person says, "No, I don't agree it's raining."

Do you automatically label the person as delusional and therefore discount their dissenting vote? No. Not yet, anyway. You need to determine where his reasoning departs from yours. Although a statement like, *"It's raining,"* seems to be digital in nature, (that is, either it's raining or it isn't), in reality, even the simplest claim contains many components. I would turn to the dissenter, "Do you agree with the definition of rain as *condensed moisture from the atmosphere which falls visibly in separate drops?"*

"That seems to be the mainstream definition," the dissenter says.

"Okay. And do you agree that currently there are separate drops of moisture falling all around us?"

"Certainly," the dissenter says, extending his hand from beneath his umbrella. "But I'm not convinced that these drops of water are the result of *condensed moisture of the atmosphere*."

Ah-ha! We've found where his reasoning departs from ours. So let's ask, "What the heck else could these water droplets be?"

We've cornered him. Now he has to put his cards on the table. "I believe in cloud gremlins. They live in clouds and are wringing out their wet towels. That's where this water is actually coming from. I'd show them to you, but they're invisible." (*)

When someone contradicts the consensus theory (a vital part of scientific progress, by the way), we need to hone in on the point of contention and address that particular claim. Whether we're discussing the theory of rain, or the theory of evolution, you need to have them admit their specific, divergent belief. Then simply ask them to produce evidence that...

- Invisible cloud gremlins create raindrops.

- An invisible cosmic consciousness somehow guides evolution.

...etc. We put the burden on *them* to prove their competing theory.

As the dissenter, if you're someone like Galileo and you've got the evidence to support your claim, then that's awesome! We were all wrong, and it's your name that gets etched on science buildings and on space probes. But if you don't have the evidence, then you should reexamine your belief and ask an expert to explain to you how moisture condenses out of the air, or how small changes over time can result in extremely complex organisms, and so on.

(*) <u>Footnote</u>
Reading that, a believer will laugh, "*Invisible cloud gremlins?* That's ridiculous!" all the while believing in some invisible supernatural being to explain some process that they themselves don't understand, be it evolution or the expansion of the universe.

CHAPTER 5:
The Simplest Explanation
is Always Preferred

INTRODUCTION:

If you thought some of the claims in Chapter 4 were incredible, wait 'til you see the ones in Chapter 5! Area 51, Bigfoot, Astrology, Psychics and the Afterlife are just some of the claims we'll discuss. Remember, I won't be doing much debunking of these claims. That territory has long been covered. Instead, as we look at and discuss these beliefs, what I hope to point out is what I call a *commonality of complexity*. As you'll see, the kinds of complicated explanations we encounter again and again are inevitably a sign of human fabrication.

STATEMENT: Seventy-six million years ago an intergalactic ruler named Corthu brought billions of his people to Mars, stacked them around volcanoes, and then detonated atomic bombs inside the volcanoes to destroy the people. Unfortunately, when these people were destroyed, their souls—called "betans"—flew to Earth and latched onto the humans there. It's these betans that are the cause of all human misery today.

AGREE:
I can't imagine that anyone would agree with this claim because I created it just now for this book. The statement is, however, quite similar to what Scientologists believe. (To discover the particulars of their religious tenets, simply search the net for "Scientology beliefs.") Assuming you don't actually believe the bizarre claim written above, then please go to the 'DISAGREE' section below.

146

DISAGREE:

As I mentioned above, the opening statement is similar to the fantastic story at the heart of Scientology. My goal, though, isn't to debunk the tenets of Scientology or any other religion. Instead, if you're a believer, my hope is that you might objectively reexamine your belief in them. And that's where the thesis of this chapter comes into play: *The simplest explanation is always preferred.*

The claim about Corthu and the betans, along with the similarly fantastic claims of Scientology, are both potential explanations for human misery. They're certainly imaginative. Still, one can't help but wonder: ***Isn't there a simpler explanation for human misery?*** Perhaps our woes are simply the result of our needs and wants not being fulfilled. For example, I need shelter from the elements, and I need food. If I'm out in the freezing rain and haven't eaten all day, I'm likely to feel miserable. I see no need to posit the existence of ancient, intergalactic betan souls for that, do you?

When stated openly, without the buildup they likely receive when presented in church, the beliefs of Scientology come across as a bit...***unrealistic***. But are they really any more fanciful than most other religions? I remember reciting the actual beliefs of Scientology to a Christian friend of mine, and he rolled his eyes. "A galactic **warlord?**" he mocked. "Souls coming out of bombed volcanoes? Gimme a break!"

"Yeah, it's pretty crazy stuff," I agreed. "Say, remind me again what you believe in?" Now, he didn't phrase it exactly as I have it below, but nevertheless this is what he said:

"I believe," my Christian friend said, "that a young woman magically got pregnant without having sex. Her child had magical powers like being able to walk on water and raise people from the dead. When her son was executed, he magically came back to life, and then went up into the sky to be with his invisible father. He did this to make up for all the bad things people do. At some unspecified point in the future, he will return to Earth surrounded by a lightning storm and heralded by magical flying creatures with trumpets. Then, only those who previously had declared a belief in him will be taken to a magical plane of existence to live forever."

When we look at the big picture, it's not just religious claims that seem unnecessarily complicated. *Conspiracy claims* like the ones regarding the attacks of September 11th, or the purported moon landing hoax are also stupefyingly complex. *Origin myths*, too, are head-scratchingly complicated, whether it's the story that the Earth and sky were created when two magic turtles were fighting over a spot of mud, or a postulation that an invisible being took a bone from a man's chest and magically created a woman in the presence of a magical snake. And still more complicated explanations are necessitated by all manner of *paranormal claims*, from alien crop circles to ESP.

To me, this recurring theme of elaboration is revealing. When we lack information or lack an ability to *understand* the information we have, (or both!), our brains inevitably weave a much more elaborate quilt from the patches of information we do grasp. The stories end up being complicated because there's an infinite amount of incorrect explanations, most of which are going to be far more involved than the *actual* explanation. It's not particularly profound to say that when we don't understand something we make stuff up. But by noticing this *commonality of complexity* amongst all these types of claims, it gives us the confidence to cast a skeptical eye at the incredible claims we might encounter in the future.

STATEMENT: There are certain psychics known as mediums who can communicate with dead people.

AGREE:
They have ordinary names like John Edwards, Sally Morgan, James Van Praagh, Rosemary Altea, and Sylvia Browne, and yet the powers these television "mediums" lay claim to are of an extraordinary nature. Assuming you truly believe that these so-called psychics are capable of talking to dead people, then I'm hoping I can get you to reexamine that belief.

We need to start by asking what reason is there to believe that? That was our takeaway from Chapter 3: **What mystery exists for which *"He is communicating with a dead person"* is the only possible explanation?** Is it something like the mysterious situation below?

> James Van Praagh asks an audience of 200
> people if anyone can connect with the name
> 'John.' Sure enough, a woman raises her
> hand and says that was her grandfather's
> name.

A believer watches that and thinks, "Oh my Lord, he just connected with that woman's grandfather!" Now, I know it's comforting to think there are other realms and spirits out there. It gives us hope that this brief life isn't all we get. But don't you want to be certain? Don't you want clear cut, undeniable proof before diving into the murky waters of magical thinking? If so, then ask yourself what might have happened if you'd randomly chosen a similarly common name and asked the audience the same thing: "*Can anyone connect with the name David?*" Don't you think someone in an audience of 200 people is likely going to be named David, or knows someone named David? Instead of the "medium" being in tune with mystical forces, isn't it more likely he's in tune with a list of common names?

Then ask yourself, "**If a dead person is speaking to the 'medium,' why does the 'medium' have to address the whole crowd in such a broad, unspecific manner?** Wouldn't an actual spirit tell him, "See that lady in the red sweater, in the third row? Her name is Sandi, with an 'i'. Sandi Zimmerman. Tell her you're communicating with me, her dead father Morty. To prove it to her, have her ask me—I mean *you*—any question about our trip to Cape Cod when she was twelve."

Actually, some psychics investigate their clients ahead of time, so as to be prepared with that kind of specific information. And in these live shows, who knows what prep work is done beforehand. I suspect that some of them hand out blank cards to the people as they wait in line, to get specific tidbits of information from them, if only name, age, and address...plus who they hope to contact. They also probably have "ears" throughout the backstage area. Audience members no doubt talk to each other. How hard is it to eavesdrop on that, with all the equipment of a TV studio at your disposal? (Says one hopeful audience member to another, "Yeah, my father died a few years ago from a heart attack. I hope he's here, you know?") Some "mediums" will even circulate amongst the audience members before the taping and engage in seemingly casual chat about why they're attending. So, on the rare occasion a "medium" gets a specific hit, inevitably there was some kind of information obtained beforehand.

But remember the Golden Rule of paranormal investigation: **My inability to explain some phenomenon only means my investigation skills are limited. It does NOT count in any way as proof of your claim.** Even if we can not conceive of any way that the "medium" could've gotten the information by natural means, that still isn't proof that *their* supernatural explanation is the correct one! The following are equally valid paranormal explanations:

* Zeus (who is an all-knowing god) is communicating the information about the audience members to the psychic.

* Invisible leprechauns follow each of us around, and **they** are the ones whispering in the psychic's ear.

* Mischievous aliens are broadcasting the answers into the "medium's" head.

I've met people who are convinced about the above alien explanation. They informed me that these aliens have been monitoring humans for thousands of years, going back to when they helped us build the pyramids. They love to abduct people and do tests on them, so this is probably just another experiment they're doing. (The aliens want to see if they can make people believe in an afterlife.) So, they use their mind probes to get the detailed information they need from an audience member, and then zap the information into the psychic's mind.

"Oh, come on! Aliens and invisible leprechauns? How ridiculous!" ...says the person who believes in talking dead people.

Here's the deal: You can't just make an incredible claim like "Dead people are communicating with me!" unless you can prove that it really is dead people, and not Zeus, nor mischievous aliens, nor invisible leprechauns.

151

Ponder that for a moment.

What test would you conduct to determine that it really is a dead human being that is communicating with the medium? The psychic might say, "I *know* it's dead humans beings who are talking because they're *telling* me who they are." But that's not good enough. The actual communicator—for example, the tricky aliens—could simply be lying to the psychic. How do we differentiate between the communications from dead human beings, and the communications from sneaky aliens? You could ask a psychic detailed questions about any dead person, and he could get all the questions right every time. That *still* is not proof in the slightest that *dead people* are feeding him the information. He could be getting the information from Zeus, or Allah, or the magical inhabitants of Atlantis who are beaming it to him with their advanced mind-machines. (Though it'd be much more likely that the person is perpetrating a clever hoax.)

Until you can prove that:

1. The spirits of dead people exist.

2. The spirits are capable of communicating silently with certain people.

3. It's specifically the *spirits* that are communicating to the "medium," and not mischievous aliens who are sending the information, nor the advanced inhabitants of Atlantis, nor invisible leprechauns, etc.

...then you shouldn't make any claims about the matter.

When it comes to explaining the things mediums do, instead of invoking the immensely complex idea of spiritual information transfer, why not search for a simpler (albeit less intriguing) explanation for how they do their readings? Perhaps they're simply using a combination of:

* **cold reading** (where they make lots of vague guesses and pounce on anything that's confirmed)

* **warm reading** (where they say things that are true of everyone)

* and **hot reading** (where they gather information from the crowd ahead of time).

To really get a feel for these techniques, take a minute to play the role of "medium" yourself. Pretend you've been asked to stand in for John Edwards. How would you start? Below is my version, which demonstrates both cold and warm readings. I start by addressing my audience of 200 people...

PSYCHIC: I'm feeling the presence of a male energy. Does anyone have a connection to a man who has passed? It could be your father, an uncle, a friend of the family, your grandfather, your great grandfather...

(Approximately 197 people are nodding their heads, so the psychic begins to narrow things down a bit.)

PSYCHIC: Does anyone have a Scorpio connection that's in some way related to that person? It could be a Gemini connection, too. Or a Sagittarius? How about Leo, or a Capricorn or a Virgo or a....?

(A woman in the audience raises her hand...)

AUDIENCE MEMBER: My father passed away, and he had a *friend* who was a Capricorn.

PSYCHIC: Yes, I know. Well, they're telling me to ask you about a name that has the letter J, R, T, or S in it somewhere. Or perhaps an M, N, P, or D? Any of those letters make sense? Or any name with an O, E, A, or Y in it? Can you connect the name of anyone you know, living or dead, to any of these letters?

AUDIENCE MEMBER: Umm...

PSYCHIC: They're not saying the name has to *begin* with one of those letters. It just has to be in the person's first or last name somewhere.

AUDIENCE MEMBER: Nothing yet.

PSYCHIC: ...or be *alphabetically near* one of those letters.

AUDIENCE MEMBER: Yes! Remarkable! My father's name was Bill!

PSYCHIC: Yes. "Bill." That's the name I'm hearing. And I'm feeling that his passing was due to either some kind of *accident*....?

AUDIENCE MEMBER: No.

PSYCHIC: ....or sickness?

AUDIENCE MEMBER: Yes, you got it exactly! This is incredible!

PSYCHIC: Right. Well, he's telling me that his illness made him very sick. And I'm feeling it was something in the brain, is that right? Or his chest somewhere? Like, the heart or his lungs, maybe? Or in his digestive tract or colon, it could be? Anything?

AUDIENCE MEMBER: Not just yet.

PSYCHIC: ...or some organ kinda *near* any of those is what I'm feeling.

AUDIENCE MEMBER: Yes, exactly! He died of pancreatic cancer.

PSYCHIC: Right, pancreatic cancer. That's what it was. Well, he's saying that he loves you and misses you. And he says don't worry about that time you did something wrong as a kid. He says you know what he's talking about.

AUDIENCE MEMBER: *(stifling a tear and nodding)* Papa!

PSYCHIC: *(to the applauding audience)* Heartwarming, isn't it, when a connection is made?

AUDIENCE MEMBER: Say, while you're communicating with him, could you ask my dad what the combination is to the safe in his office? We can't get the darn thing open.

PSYCHIC: Oh, I'm so sorry. Your father has moved on now.

* * *

To any rational thinker, the simple explanation of cold, warm and hot readings should be preferred over the vastly more complicated supernatural alternatives.

DISAGREE:
If you read the 'AGREE' section above, you may have noticed that I put the word "medium" in quotes. I did this because it's not a real word. It was coined by our gullible and non-scientific Victorian-era forebears and doesn't describe any real ability. If we say, "A medium is someone who can interface with the spirits in other dimensions" (to quote Mr. Van Praagh) it's a nonsense definition. It's very much like saying a "leprevoyant" is someone who communicates with leprechauns. Or, a "gremlinum" is someone who can communicate with gremlins. Leprechauns and gremlins and spirits don't exist, so why do we need words for those who pretend to communicate with them?

At the very least, the entry in all dictionaries should be changed to read:

> **medium**: Someone who, resulting from
> delusion or an intent to defraud, claims to
> communicate with non-existent lifeforms.

What's ironic is that the believers accuse the skeptics of not having an open mind. The reality, of course, is just the opposite. I am willing and eager to believe anything, given sufficient proof. Believers, on the other hand, have made up their minds and nothing will ever sway them. As for psychic powers, perhaps one day researchers will prove that some people are capable of receiving information in some novel way, not connected to one of the five senses. That'd be great! And assuming the experiments passed a stringent peer-review process, I'd be the first to acknowledge this new 6th sense. I admit I find it extremely unlikely, but nevertheless if such a sense is one day discovered, it would likely turn out to be some receptor in the brain which is sensitive to some kinds of real-world signals. No big deal. The supernatural would become natural.

STATEMENT: The U.S. military is keeping the remains of an alien craft (which crashed near Roswell, New Mexico) at a secret base called Area 51.

AGREE:

What actually fell from the sky near Roswell that day back in 1947 was an experimental high-altitude surveillance balloon. There have been other crashes in the vicinity after that date, including one involving badly burned (human) pilots. Over time, the stories got jumbled and were elaborated. This is the rational explanation for the Roswell phenomenon. Notice how it's orders of magnitude less involved than the "alien crash + government coverup" conspiracy theories which have been fabricated decades later regarding the Roswell events. Of course, I realize I'm talking to a Roswell/Area 51 believer here, so you've already heard (and dismissed) the surveillance balloon explanation. Nevertheless, you're here reading this, hopefully with an open mind, so maybe you'll read a bit further. If so, I have some questions for you:

How did you first come to believe that an alien craft crashed near Roswell?

Were you on the Foster ranch that day and thus you clearly saw the craft and its occupants? Were you one of the Air Force workers who handled the craft or its occupants back at the base? (If you were indeed there, then how do you know the debris was of an alien craft, and not, say, a Russian spy craft of some sort?)

How do you know the occupants were specifically extraterrestrials?

Mr. Glenn Davis (the main witness of "alien bodies", whose testimony is a pillar of the Roswell belief) surely did see bodies. *Human* ones. Eleven Air Force pilots died and were horrendously burned in a terrible crash. (Yes, I know. That's just what the government wants me to believe.) But my question is: Even if you're convinced they were not human, why do you make the huge leap and declare that they are *extraterrestrial* bodies? Perhaps they were from the advanced civilization of Atlantis, here on Earth. Perhaps they were humans from the future, traveling back through time. How do you know with such certainty they were aliens? Will you at least be honest and admit that *that part* of your claim is speculation?

Think of it this way: Pretend the Air Force lets you in to see the bodies. How would you prove they were extraterrestrial, and not from Atlantis, or not future humans? Even if they had no DNA, they could still be creatures that we engineered. **How would you prove they were not from Earth?**

Why not take the U.S. Air Force to court?

You have all this "proof" about the existence of an alien craft and alien bodies. The evidence must be utterly overwhelming, or else why would you believe it so fervently? Why not sue the U.S. Air Force? Anyone seeing all your incontrovertible evidence could only draw the same conclusion that you did, right? So it should be an absolute breeze for you in court. Lawyers love these high profile cases, especially when they're easy to win, so you shouldn't have any problem finding an eager lawyer. So what's stopping you?

Have you honestly taken the time to carefully research the answers to your Roswell questions?
History can be tough to piece together to form a coherent narrative. There were multiple crashes, over the course of many years, in different locations. Different witnesses reported different things, and it's hard to separate hearsay from testimony, and to recognize reliable memories from unreliable ones. When you ask questions like...

What could have happened so long ago at Roswell to require such high levels of security?

Why were trained officers so puzzled about the debris that was found?

...what you're really saying is:

I'm not smart enough to understand how important security is to military operations.

I'm not smart enough to realize that it might be difficult for a soldier to know what an advanced piece of military equipment would look like after it's been mangled in a free fall crash.

Instead of broadcasting your ignorance, you should be reading scientific explanations about the things you don't understand.

Why don't you ever ask similar questions of your own theory?
It's good to ask questions like the ones above, but you need to be consistent. Why do you not take aim at your own theory? Shouldn't you also be asking:

* If all these soldiers are willing to tell the truth about the aliens they feel they saw, why wouldn't they show us a piece of the craft, or some miniscule tissue sample, to really document their claim? They'd be vindicated, they'd be heroes, and they'd be rich.

* These aliens have the ability to cross the trillions of miles of interstellar space, and then they clumsily crash when they finally get here? And worse, they had no backup plan? They had no support and rescue team?

* Do you really think aliens would come all this way in secrecy, only to leave their *lights on* for us to see? (It's not a question directly related to the Roswell/Area 51 conspiracy, but I've always wanted to ask it of UFO enthusiasts.)

You don't ask these kinds of questions about your own theory because you're not interested in the truth. Instead, as with all conspiracy theorists, you're interested in feeling like you have secret knowledge that few others do.

Don't you realize what you're missing out on?
It might feel cool thinking you're in on some secret that no one else knows about, but it's not. It's sad, because you're missing out on something wonderful: *Life*. Every day you waste chasing a fantasy could've been spent learning more about reality. However cool your imaginary alien craft is, the real things going on in science today are infinitely more fascinating.

DISAGREE:
You have to admit, the idea is secretly very pleasing. An incredible new reality of alien civilizations lies tantalizingly close, just beyond the guarded perimeter of a nearby military base. If not for our tyrannical government, we, too, could share in the wonder of our extraterrestrial visitors. The irony is, I would love that more than any of the muddled thinkers who buy in to the Roswell conspiracy. I would literally weep if it were proven true that extraterrestrials exist and have visited Earth. But there is ***not one molecule of evidence*** that would make any rational person believe that extraterrestrials have ever visited Earth.

So why do some people latch on to an utterly unfounded claim? As we learned when discussing both the moon hoax conspiracy theory and the 9/11 conspiracy theory, the Roswell believers are motivated by two simple things: **increasing pleasure** and **minimizing pain**. It feels great knowing you're privy to Something Very Important that the rest of world is in the dark about. That excitement is always the lure that reels them in. But despite all the rational answers to their litany of questions, they persist in their belief to avoid the pain of acknowledging they made a mistake. They'd have to say, "Wow! I have no critical thinking skills. I've been wrong all this time, and have been making a fool out of myself in front of my more skeptical and rational friends. How embarrassing." It's much less painful to dig in more deeply and shoot down all the explanations which undermine their claim.

RESOURCES

Here's a great article from the New York Times explaining the Area 51 stories. *Air Force Details a New Theory in U.F.O. Case: A Suggestion That Dead 'Aliens' Were Test Dummies* By William J. Broad. It's available for free, online: <http://www.nytimes.com/learning/general/onthisday/big/0624. html#article>

STATEMENT: In the forests of the Pacific northwest lives an undiscovered primate larger than man.

AGREE:

I could've written the above statement in just two words: "Bigfoot exists," but I prefer to restate the claims people make using more precise terminology. It forces them to re-think their bizarre beliefs, if only for a moment. "Oh yeah, that *is* what I believe. Hmm..."

Before going any further, I'm going to assume that since you agree with the statement about the existence of an undiscovered primate in the Pacific northwest, then you also agree with this:

> **STATEMENT**: In a deep lake in northwest Scotland lives an undiscovered, dinosaur-like creature.

I'll admit it: I believed both of these when I was a boy. For the existence of Bigfoot, I based my belief on the Patterson film. (What else *is* there to base it on?) And yet in all the time since its release (45 years and counting), not so much as an

162

anomalous hair sample has ever been discovered, to say nothing of an anomalous *body*. You have all these cryptozoologists (as these monster hunters call themselves) searching high and low, decade after decade, and not a single carcass or bone from a deceased Bigfoot has ever been found. ***Do these creatures magically disintegrate when they die?*** The bottom line is: There's just no reason whatsoever to believe such creatures exist.

Ditto for Nessie. In fact, even *less* reason exists to believe that dinosaurs roam the depths of Loch Ness for the obvious reason that dinosaurs went extinct 65 million years ago, and pleisosaurs in particular had disappeared another 100 million years before that. Add to that the fact that the one clear photo of the supposed monster, the infamous "Surgeon's Photo" was admitted to be a hoax. But that's just nitpicking. The fact is, just like with Bigfoot, not a single bit of biological evidence exists that might lead us to believe that the supposed dino-creature is anything but the misinterpretation of imperfect sensory input, and of course, hoaxes.

Remember the mantra of rational thought that we've been chanting throughout this chapter: ***The simplest answer that accounts for all the data is to be preferred.*** In this case, all accounts of the supposed creatures can be explained by imperfect observations, faulty interpretations, and hoaxes. That is a much simpler explanation which accounts for all the data. In comparison, the idea that *hundreds* of these large creatures (there have to be enough members to sustain their populations) could remain undiscovered in confined locations despite all efforts to find them? That requires a vastly more complicated explanation. I'm not arguing from ignorance. No one is saying it's impossible. The planet is a big place and we've discovered big creatures before. But there's simply no evidence commensurate with these particular claims.

DISAGREE:

It can be depressing to let go of a cool belief like Bigfoot or the Loch Ness monster. The good news is, there's a much greater thrill to be found in following the very real discoveries of paleontology. Have you read about *Cronopio dentiacutus*, the Saber-toothed squirrels discovered in Argentina? How about Pampaphoneus biccai, aka the "Pampas Killer," which looks like a cross between a tiger and a Komodo dragon? Or my favorite, *Homo floresiensis*, which is evidently a new species from the genus *Homo* who averaged about three feet in height. (I say they're **evidently** a new species because the debates between paleontologists can continue for years, even decades, before a consensus view is formed.) The amazing thing about this potential species is that they apparently survived on the Indonesian island of Flores right up until about 12,000 years ago. In other words, this hobbit species co-existed with modern humans. Can you imagine?

Of course, we're here to talk about belief, not paleontology, which is why I mentioned *Homo floresiensis*. Right now, my belief is suspended. It seems that the evidence is indeed indicating they're a new species, but what do *I* know? I'm not an expert. All I can do is follow the debate and weigh the arguments as best I can. It's fun to imagine a humanoid species living side-by-side with us. Perhaps they still exist today in the unexplored depths of the Indonesian rain forest. But I'm most interested in the truth. If it turns out they're simply a group of pygmy humans, then so be it. As go the facts, so go my beliefs.

STATEMENT: In response to prayer, God will sometimes cure serious injuries and illnesses.

AGREE:

I have many Christian friends who feel that God helped them through a variety of difficult times in their life. As a believer, you, too have probably experienced God's direct influence on the course of your life. That's not to say that God answers all prayers, but to believers there is no doubt He does sometimes respond positively.

The question at hand, though, is whether this supernatural being occasionally responds to pleas for help by ***miraculously healing people's injuries or illnesses***. A large percent of the population seems to believe so. Take a look at any social networking site like Facebook, and you'll find posts from people offering to pray for someone else's loved one who is sick or injured. If you truly believe that God will sometimes cure serious injuries and illnesses in response to prayer, then the question you need to ask yourself is this:

Why won't God ever heal an amputee?

Think of all the American and allied soldiers—many of whom are Christian—who have lost a limb fighting in Iraq or Afghanistan. If anyone deserves to have a limb regrown, it's these incredibly brave, selfless young soldiers who put their lives at risk for our safety and freedom. Why won't God regrow their limbs?

[**NOTE**: This is the topic of a website appropriately titled: www.WhyWontGodHealAmputees.com, which was my original source of this argument against prayer. Although the words here are my own, the basic argument is from their site, and the articles there are a great source of rational discourse on the topic.]

Imagine logging on to Facebook and announcing, "Hey everyone, **please ask God to regrow my cousin's legs** which were blown off when a bomb exploded beneath him in Kabul." Even the most devout Christians would think, "Grow back limbs? Are you kidding me? It doesn't work that way." Which is, of course, true. It definitely does not work that way. We all know that human limbs don't grow back. And we all know that no matter how deserving the person is, no matter who was praying, and no matter how long or fervently they prayed, that God will not regrow the person's limb. (Which seems particularly cruel and petty considering he'll grow them back for a lowly salamander or starfish.)

Obviously, the only ailments this "all-powerful" God can *actually* fix are things that can get better naturally. Have a fever? Stub your toe? Then prayer works wonders. But if you've got, say, a severed spinal cord, don't waste your breath. He is clearly incapable of helping.

You'd think a believer would say, "Man, I never thought about that. It really is a good point. I still believe in God, but it does seem that praying to Him is ineffective." And yet that never seems to happen. In fact, it's amazing to me the lengths a Christian will go to rationalize this glaring discrepancy:

166

God has other plans for amputees.

If God regrew limbs then everyone would believe in Him.

Amputees are His testament to sin.

...and so on.

They vigorously defend God for the same reasons that conspiracy believers so tenaciously defend their own beliefs: To maintain the pleasure of believing, and to avoid the pain of admitting an embarrassing, lifelong mistake.

This chapter, though, is about our preference for simpler answers. So let's rephrase our topic statement as a question: **How does prayer work?**

The Christian Answer
There's an invisible, otherworldly supernatural being that reads everyone's mind and sometimes cures people, but only when it's something that could have gotten better on its own (like cancer) and never something that is incurable (like Lou Gehrig's disease, Down syndrome, or amputated limbs).

The Rational Answer
Sometimes certain diseases go into remission.

Notice how complicated the Christian answer is, and how simple the rational one is.

But there's a bigger question at the heart of this issue about why God won't heal amputees, or heal anyone who suffers from an incurable disease: **Why would an all-loving God allow suffering?**

There isn't one universally agreed upon Christian response. Nevertheless, the general Christian argument for the vast amounts of sorrow that exist in the world is as follows:

The Christian Answer
When God first created man upon the Earth, suffering and sorrow did not exist. But then Adam decided to break God's law by choosing the way of Satan, who is an ungodly angel that God created. Upset that Adam choose the temptations of Satan, God decided that humans now need to experience pain and suffering. The reason God decides to torture even the youngest and most innocent children by starving them, killing them in car crashes and war and so on is because even babies are sinners. And he is showing us that something is wrong between man and God, and even these children that God is killing should have come to Him and accepted Christ as their savior.

The Rational Answer:
There is no God.

Again, notice how extraordinarily complicated the Christian answer is. That's our clue that it's a human fabrication and not a factual answer. It's just like the incredibly elaborate and yet muddled explanations for how and why our government executed the attacks of September 11th, or how the crashed surveillance balloon near Roswell was actually an intergalactic spaceship which the government is hiding from us. **The simplest answer that accounts for all the data is the best choice**.

* * * *

As a closing note on the topic of prayer, I can't help but ask:

168

Why would you need to pray to an omniscient being? He already knows *everything*, and that includes knowing what you want. In fact, he knew you were going to want it before you decided to ask. Praying to God is actually an insult to Him! **You obviously feel there's something he doesn't know, otherwise you wouldn't tell him.** Each time you pray, you're showing him you don't believe he's all-knowing. And worse, you're questioning his divine plan! You're asking your perfect God to change His inherently perfect plan. It's selfish and insulting.

DISAGREE:
Despite the arguments laid out in the 'AGREE' section above, many believers still insist that God healed them or their loved one. These most stubborn believers need to be reminded of the following: **Even if you indeed recovered miraculously after praying—even if your amputated limb actually grew back! —it *still* wouldn't prove that your God did it.** There are all sorts of equally valid supernatural possibilities:

* It could be that an alien (perhaps one of the ones that built the pyramids) was eavesdropping on your thoughts and took mercy on you. He abducted you one night and healed you.

* It could be that the advanced and generous inhabitants of Atlantis were using their mind ray machine to hear your prayer and so they sent you their magical healing rays.

* Perhaps magical invisible elves follow each of us around and sometimes decide to cure us.

* Maybe we're all just avatars in an advanced computer program and the system's administrator decided to heal you.

A Christian will respond that he "knows in his heart" that God healed him, but he is only lying to himself. The aliens who *actually* cured him (or the inhabitants of Atlantis who did it, or the invisible healing elves, etc.) were simply covering their tracks when *they made him feel in his heart that it was God*.

I have used this exact line of reasoning with some Christian friends, and they of course roll their eyes at the mention of benevolent aliens or inhabitants of Atlantis. "Don't be ridiculous!" they tell me. "Aliens? Invisible elves? You're being condescending and insulting!" Evidently, the very real concept of advanced aliens is ridiculous and insulting, but the idea of an all-powerful, invisible sky fairy that insists on starving and maiming millions of innocent children because 6,000 years ago someone named Adam ate an apple from a talking snake...that makes perfect sense.

STATEMENT: The particular alignment of celestial bodies on the day you were born has an impact on the course of your life.

AGREE:
This is the general claim made by those who believe in astrology. As you know by now, it's not my goal to debunk irrational claims. Instead, I only hope to get you to reconsider your belief in them. To that end, imagine you're talking to someone who's skeptical (but still open-minded) about astrology. They just need a little evidence. What would you do to show them that astrology works?

It's not easy to prove, is it? As I pondered this, I came up with a simple, two-step astrology experiment:

> 1. Cut out the twelve horoscope predictions published in the previous day's newspaper, but cover up the actual astrological sign that each prediction relates to.

> 2. Post the twelve statements somewhere and ask people to choose the one that describes the day they had yesterday.

If astrology is effective, then only Capricorns should choose the prediction that was made for Capricorns. Only Scorpios should choose the Scorpio prediction, and so on. I would repeat this with various newspapers, being sure to use only professional, experienced "astrologers."

Imagine the above experiment. What do you think the actual results would be? Personally, I suspect the results would be that for each of the twelve predictions, you would have an equal assortment of all twelve astrological signs. That is, when you revealed the prediction which was intended only for Virgos, you would nevertheless see people from all twelve signs saying that the prediction applied to them. And so it would be for each of the twelve astrological predictions. Doesn't that seem like the inevitable result? The question then becomes *why*? Why would a prediction designed just for one astrological sign actually attract all people to it?

Okay, so it seems the predictions made in newspapers are purposefully vague so as to fit all readers. But perhaps you believe in the more personalized astrological readings that are worked out by professional "astrologers" based on the exact moment of a person's birth. Again I must ask: What test would you devise to prove to that such astrology charts offer useable information? Here's the kind of experiment I'd like to see:

> 1. Twelve people submit their birth information to an astrologer.
>
> 2. The astrologer writes out specific predictions (or statements) for each and sends each person all twelve of the unlabeled readings.
>
> 3. Each participant then chooses the one reading that relates specifically to them.

If the astrologer is actually in-tune with a higher plane of cosmic knowledge (whatever that means), then the Virgo

172

should choose only the reading intended for him and be utterly repulsed by the blatant inaccuracies of the other eleven readings. So, too, should all the other participants. They should each immediately see the clear truth written in the reading intended for them, and quickly pass on all the other readings. In fact, if you filmed their faces as they were looking at each of the wrong readings, you would expect to see them cringe at the ridiculous statements being made. Only when their eyes happened upon the sweet truth of their particular reading would a smile spread across their face. "This is obviously written for and about me!"

Do you think that would happen?

Similar experiments have been conducted, and I think you can guess what the results were. Because the statements that astrologers make are always vague and can apply to anyone, their results are never any better than chance.

In this chapter I'm encouraging you to look for the simplest explanation that accounts for all the data. Surely, if ever there were a field where this principle of "simple is best" applies, it's astrology. So let's rephrase our topic statement as a question and then compare the astrologist's answer with the one given by rational people: **How does astrology work?**

The Astrologist's Answer
Both the gravitational pull of the planets, as well as the planetary energies themselves have effects on our lives because we are all connected to the cosmos through a multi-dimensional reality. Astrology, therefore, is a kind of map linking us to the expression of these extra-dimensional realities.

The Rational Answer
Astrologers make vague statements that apply to everyone.

Notice how complicated and convoluted the astrologer's answer is, and how simple and straightforward the rational one is. As always, that's our clue that astrology is a human fabrication and not a logical explanation. It's no different than the elaborate explanations that religions make for how the world came to be, nor any different than the exceedingly complicated explanation for how the moon landing "hoax" was achieved.

And please keep in mind the overarching message of the book itself: **By believing in astrology, you're missing out on something wonderful:** *self-confidence*. If you attribute all your best qualities and all your achievements to the irrelevant alignment of stars, then you're not giving yourself due credit! You're successful at what you do because you have innate talent and have worked hard, not because the planets were aligned for you. You're a good friend because you care about people and are a good listener, not because "Pluto was in the third house." And you'll continue to make the best of your opportunities in life not because some "astrologer" says so, but because you're an earnest, goal-oriented person.

DISAGREE:
To quote the sublime science fiction writer Arthur C. Clarke: "I don't believe in astrology; I'm a Sagittarius and we're skeptical." That quote makes me laugh every time, and indeed the whole subject of astrology would be laughable if these hucksters weren't taking hard-earned money from their followers.

My real issue with astrology is that there's simply no unexplained phenomenon that necessitates a belief in it. I mean, what is the anomalous data? Are these astrologers making specific predictions that are then coming true? If so, it should be a very easy thing to prove. I'd like to do a strict trial to test their ability to predict random events. For example, based on my birthday, they should be able to predict the numbers I'll get when I throw a die repeatedly: "According to your chart, you'll throw a 5, then a 2, then another 5, then a 1, then another 1...." But when you corner an astrologer, they say that they don't make predictions of any sort. They only give general advice. My response: *So, how does that make you any more useful than my grandmother?*

If you read the 'AGREE' section above, you may have noticed that I occasionally put the word "astrologer" in quotes. I did this because it's not a real word. Like the equally vacuous word "medium," it was coined by our gullible and non-scientific ancestors and doesn't describe any actual ability. If we say, "An astrologer is someone who can deduce the effects of the planetary energies through different planes of existence," (to quote one astrologer's website) it's an utterly nonsensical definition. It's very much like saying a "mudologer" is someone who gives advice based on the reading of mud cracks. Or, a "cloudologer" is someone who gives advice based on the shapes of clouds. Astrologers aren't giving any measurably significant data, so why do we need words for those who pretend to do so? I feel that the entry in all dictionaries should be changed to read:

> *astrologer*: Someone who, resulting from delusion or an intent to defraud, claims to offer specific advice based on their interpretation of irrelevant celestial information.

Shouldn't language strive to be accurate?

STATEMENT: The descriptions from people who've had near death experiences are proof that there is life after death.

AGREE:

The tunnel, the bright light, the sense of floating above their own body...we all know the elements of the NDE, or "near death experience." Thousands of people who've survived drownings, heart attacks, violent accidents and other traumas have recounted these same sensations. *They can't all be lying,* insists the believer. *There must be something to it.*

I agree. There indeed must be a real phenomenon at work here, which definitely warrants an explanation. But how does the data of NDEs lead us to conclude that there is life after death?

To be honest, it's completely nonsensical. Do you have a vastly different definition of "life" than the rest of us? Life is a process exhibited by physical entities whereby they react to stimuli, take in energy to maintain their stasis, and eventually reproduce. ***How can you claim there is life after death when you can't even produce the physical entity you are claiming is alive?*** Put another way: I'll be glad to believe there's life after death, just as soon as you show me the thing that's living.

The focus of this chapter has been to encourage you to choose the simplest explanation for extraordinary claims. To do that, we need to rephrase the claim as a question: **Why do so many survivors of traumatic, near-death experiences recount such similar experiences about a tunnel, a bright light, and a sense of floating?**

The Believer's Answer

Human beings have an invisible, supernatural "spirit" that exists on another plane and is therefore impossible to detect and yet it magically contains all the information stored in our brain and is able to actively process that information, as well as magically record sensory data despite the lack of any sense organs. This invisible, undetectable "spirit" detaches from the body when death is imminent and travels along a tunnel towards a light (both of which are, of course, invisible and undetectable). This light is actually a manifestation of the invisible, undetectable divine creator.

The Rational Answer

When the brain is deprived of oxygen for prolonged periods, a set of physiological events occur which comprise the NDE experience.

Notice how complicated the believer's answer is, and how simple the rational one is. For a thorough explanation behind the entire NDE experience, I recommend Michael Shermer's book The Believing Brain (pg. 141).

DISAGREE:

The British philosopher Chris Carter claims that the mind exists outside of the brain. With that as his fundamental premise, he wrote the book <u>Science and the Near-Death Experience: How Consciousness Survives Death</u>. How a philosopher with a degree in economics is qualified to write a "science" book escapes me. One claim he makes is that the Neanderthals believed in an afterlife. Really, Chris? How exactly did you leap to that wild conclusion? Did you consider that perhaps they buried their dead to avoid attracting predators? And even if Neanderthals did believe in an afterlife...so what? How are Neanderthal beliefs relevant? It's like claiming the earth is flat because that's what the Neanderthals believed.

In an interview with Alex Tsakiris on a site called skeptico.com in December, 2010, Chris disingenuously claims, "The rival hypothesis that the brain produces the mind has thus been proven false by the data." (Carter)

Rival hypothesis? Sorry, Chris. The statement that the brain produces the mind is the *only* working hypothesis. That is obvious to the point of being self-evident. And to have the further gall to declare it's been proven false? Are you kidding me? Such a statement lowers Mr. Carter to the level of Flat-Earthers who claim the "rival theory of the round Earth has been proven false."

In the interview he goes on to say, "I've tried to explain that there's no good reason, either from philosophy, from science, from common sense, to think that the brain produces the mind." Why would Chris "try to explain" anything, if he just told us his claim has been proven? His lack of logic aside, it's glaringly wishful thinking on his part to say there's "no good reason from science or from common sense to think the brain produces the mind." Statements like these, of course, show us why Chris is a mere philosopher and not an actual scientist.

Meanwhile, neither Chris nor anyone else who claims that the mind exists outside of the brain have submitted a testable, *falsifiable* hypothesis for their supernatural explanation behind the NDE phenomenon. What, then, is Chris's theory? He, of course, doesn't really have one. The best he can come up with is an unfounded analogy that our brains are merely like television sets, receiving the signal of the mind from another magical, undetectable plane. An unnecessarily complicated idea with no basis in reality and utterly untestable. You might as well say our minds are all located on-board Xenu's space cruiser, in his "galactic mind machine." (You remember Xenu as the evil galactic dictator from Scientology.)

If the mind exists outside the brain, then why are our mental abilities tied so directly to the condition of our brain? This is what I meant about the whole "brain causes mind" issue being self-evident. A complex human brain results in a complex human mind. A less complex chimpanzee brain results in a less complex chimpanzee mind. And if you damage a human's brain, his corresponding mental capabilities are also reduced. (That shouldn't be the case if the mind were stored safely away in a magical plane of existence.) Chris's book conveniently never addresses this obvious point.

I need to end this with the main point from the 'AGREE' section above. The believers are claiming there is life after death. That is the purest form of nonsense imaginable. Life is a process exhibited by *physical entities* whereby they react to stimuli, take in energy to maintain their stasis, and eventually reproduce. *How can you claim there is life after death when you can't even produce the physical entity you are claiming is alive?* I'll say it again: I'll be glad to believe there's life after death, just as soon as you show me the thing that's living.

CRITICAL THINKING LESSONS
FROM CHAPTER 5

Here are the takeaways from Chapter 5:

* The simplest explanation that accounts for all the data is always preferred.

* Complicated explanations are inevitably a sign of human fabrication.

PARTING THOUGHT:
I Believe in Magical Underwear

In the beginning of Chapter 5 we looked at a set of beliefs that are very similar to the bizarre science fiction beliefs of L. Ron Hubbard's religion, Scientology. But they're no more strange than the beliefs held by Mormons. As you read these, bear in mind that the religious beliefs you yourself hold are just as ridiculous sounding to the rest of us.

A Short List of Mormon Beliefs:

* It is important to wear sacred underwear which magically provide protection against evil.

* In 1823, a magical winged being direct Joseph Smith to a hill in New York where he discovered magical golden plates. The winged creature gave him magical glasses called Urim and Thummim to translate the magical golden plates, but he also used a magic stone. He placed his magic stone in a hat and pressed his face into the hat and translated the magical visions he received.

* American Indians are descendants of ancient Jews, who migrated from Israel to America sometime before Jesus was born.

* A magical place called the Garden of Eden contained an evil talking snake. In this place a magical being took a rib from a man named Adam and magically made a woman from it. This place was located in Missouri.

* Two-thousand years ago in the Middle East, a magical being impregnated a woman without sex and her magical son was crucified and then came back to life and appeared to the Indians in America soon thereafter.

* Any human who died without being a Mormon can become a Mormon through vicarious baptism.

* The invisible, magical being that created the universe lives on a magical, undetectable planet named Kolob.

(Source: LDS.org)

That's the short version but I'm sure you get the gist. Each religion is equally weird sounding if you haven't been brainwashed since childhood to accept it unquestioningly.

Finally, as an exercise in rational thought, try rewriting all the major beliefs of your own religion using more detailed, synonymous language. For example, don't use the word *angel*. Call it what it is: A humanoid creature with wings that can appear magically and transmit messages from an all-powerful being located in a magical plane of existence.

CHAPTER 6:
Is Your Hypothesis Falsifiable?

INTRODUCTION:

Although it starts out with a discussion about Thor, Chapter 6 is actually our science chapter. In it we'll be looking at claims regarding vaccinations, Martian meteorites, and global warming in an effort to understand a principle in science called "falsification."

STATEMENT: Thor is a real god, and He is the direct cause of thunderstorms and lightning.

AGREE:
Unless you're a 1,000 year old Scandinavian, you're probably reading this 'AGREE' section just for fun. But to those who believe in a god—any god—**why *not* believe in Thor?** (Read the 'DISAGREE' section to see the evidence for Thor.)

DISAGREE:
I think it's safe to say that not a single person alive today considers Thor to be God. This includes rational thinkers as well as religious believers. The common reasoning seems to be that all of Norse mythology was simply the Vikings' way of explaining things they couldn't understand. (That's certainly my own viewpoint.) Nevertheless, if you're inclined to believe in a supernatural creator, there is some very convincing evidence for Thor as God:

The Evidence for Thor:

* He and the other Norse Gods were written about in **three**
Holy texts: a pair of books called the Eddas and a third book,
the Heimskringla. These texts are the unerring word of God,
and give detailed accounts of Thor's life on Earth. (These are
ancient books dictated by God. Isn't that all one needs for
proof?)

* Thor is so holy, he has a **day of the week** named after him.
The word "Thursday" is a modification of "Thor's day." So, we
are all essentially paying homage to Thor once a week!

And the most convincing proof of Thor's power and holiness:

* All thunderstorms and lightning are created by Thor.

Whoa! What's that you say? Lightning is actually caused by the
build up of charged particles which, when the differential
reaches a certain threshold, seek to equalize in a giant spark?
And how did you come to **that** conclusion? If you're at all
religious, you need to be careful about quoting science.
Because if you're going to rely on science to explain why
there's no reason for us to believe in Thor, then you have to be
consistent and use that very same scientific reasoning to
conclude the same thing about your own god. You can't have it
both ways. Either you agree that science explains things like
lightning, evolution, and the age of the universe, (in which case
we need not postulate a supernatural being to account for
them), or you feel science is unreliable, (in which case any
magical explanation at all must be accepted.)

Of course, all believers of irrational things **pick and choose** the
science that they want to believe in. It's not just religious
believers who do that. Take a 9/11 "truther" who maintains that
the events of September 11 were a government conspiracy.
Though they accept scientific consensus in so many other
things, 9/11 "truthers" disregard the scientific explanation
regarding the weakening of steel above 400 degrees because it

violates a belief they desperately want to maintain. That's hypocritical.

Similarly, Christians will accept the consensus scientific explanation for lightning to disprove the magical idea of Thor. Then, without any shame, they disregard the consensus scientific explanation for the age of the universe because it violates a particular magical belief that they want to maintain. That's equally hypocritical.

To lay it our bare, here's a direct comparison between the way believers think, versus the way rational people think. (Again, this is true of all believers, not just religious people.)

Thought Process of a Believer:

a) I have a set of beliefs that I am heavily invested in.

b) My beliefs are the truth. They are ***not*** to be questioned!

c) When the scientific consensus confirms my belief—for example, when science says that lightning is a natural phenomenon—then I gladly accept that.

d) When scientific consensus contradicts my belief—for example, when science says that the universe is 13.6 billion years old—then I refute it.

e) As part of the refutation process, I'll support any lone dissenter who makes a claim that contradicts the consensus opinion, even when that dissenter is: 1) Not supported by his peers, and 2) Commenting outside his specific field of expertise.

f) Finally, if I can't find any recent or living authority to refute the scientific consensus, I will then tell the skeptics:

* The Bible is the ultimate authority. (Christians)

* "You've fallen for story *they* want you to believe,"
(Proponents of conspiracies about AIDS, 9/11, Area 51, the
moon landing, etc.)

* There are other planes of reality that science can not
investigate. (Believers in 'psi' phenomena such as ghosts,
mediums, life after death, astrology, etc.)

Thought Process of a Rational Person:

a) I base my beliefs on the evidence, as well as the consensus
opinion of the experts who evaluate it.

b) As more evidence comes in and the consensus view changes,
then my belief changes.

Obviously what's at issue here how we relate to science. Let's
discuss that in the next section.

STATEMENT: Science is the best tool we have to help us understand our world.

AGREE:
I should have phrased it: Science is the ***only*** tool we have that helps us understand our world. I mean, what are the alternative tools? Consulting an oracle? Reading the pages of an ancient book written by bronze-age sheep herders? If the Bible is such a great tool for understanding our world, you have to wonder why none of the Bible's authors thought to include a chart of the elements, a sketch of our solar system, or an explanation of how diseases are spread.

You and I accept the validity of the scientific method, and that puts us in the minority. Since the vast majority of people in the world are believers and not rational thinkers, there is a pervasive mistrust of science. Sure, believers will rely on science when it benefits them, but they turn their back when it casts doubt on one of their irrational beliefs. (Does no one see the irony when a Christian posts anti-science messages on an online forum? They might as well write, *"I don't believe in the very same scientific method that led to the creation of computers and the internet that I'm currently using to make this post."*)

But the same is true for other irrational believers, not just the religious ones. Take, for example, someone who believes the moon landings were all a hoax. These believers rely on science each time they drive, talk on the phone, watch TV, use electricity, etc., and yet they'll tell you, "The science is simply wrong," when it confronts their belief that the U.S. government was perpetrating the moon hoax.

Science is indeed the only tool we have to help us understand our world. All the same, even if you accept that, you might not know what a falsifiable hypothesis is, or why it's so important in science. If you're not sure, then please read the 'DISAGREE' section below.

DISAGREE:
I'm curious: Without looking it up in the dictionary or on the internet, *can you define what science is?* We hear and use the word all the time and yet it seems to mean different things to different people. And since religious believers have long been at odds with rational thinkers over the validity of science, it's probably best we define our terms, otherwise we're not engaging in any kind of meaningful conversation. So I ask again: Without looking up the answer and without asking anyone else: **What is science?**

It's not easy, is it? I've asked this exact question to many of my religious friends and acquaintances and I usually get some pretty sketchy answers. I'm paraphrasing, but they say things like:

"Science is when they do experiments and stuff."

"Science is what they do in laboratories with microscopes and all that."

I remember this next one word for word:

"Science is all the 'facts' that the scientists tell us are true."
(The quotes around 'facts' represent the sarcasm she had in her
voice. She clearly didn't consider them facts at all.)

If your own definition of science is equally vague, then it's
quite understandable why you'd take issue with labeling it as,
"the best tool we have to help us understand our world."

Think of it this way: If you've never heard of the Russian dish
"solyanka," then you'd probably not agree with the statement
that it's the most delicious food in the world. You certainly
won't give it the same praise you give to your favorite food.
Basically, you don't know what solyanka is, so you dismiss it.
That's not too big a deal when it comes to food; you're only
missing out on a very tasty Russian soup. But when you
dismiss science simply because you don't really understand
what it is, then that's much more serious. Remember, there's a
lot at stake here.

As it turns out, science is very hard to define. For one thing, the
word has different meanings:

> *science*: An organized body of knowledge
> acquired via the scientific method.

But that's not the meaning I'm after. I'm curious about the
process of science itself. How is this "body of knowledge"
acquired? Ask ten different experts what the process of science
is, and you'll get ten different definitions. Still, at the heart of
any good definition of science should be these core concepts:

> * Science seeks to understand the natural
> world and how it works, using observable
> physical evidence as the basis of that
> understanding.

> * Its conclusions are always tentative, and
> always open to modification as new
> observational evidence is compiled.

* Science can only test falsifiable
hypotheses.

"Falsifiable hypotheses?"

All that means is you have to formulate your claim in such a way so that it's possible to prove wrong. The standard example you'll see for this idea is from the philosopher Karl Popper:

"All swans are white."

That's falsifiable because you need to find just one swan that isn't white. The authors who quote this example then mention how the hypothesis was falsified when black swans were found in Australia.

My problem with this oft-cited example is this: *Define white.* If we find a swan that's "slightly off-white," does that count as "not white" and thus proves the hypothesis wrong? Perhaps I'm being petty, especially over something as insignificant as swan color, but being very specific takes on the utmost importance as the stakes get higher.

This idea of falsifiability is so crucial, it's worth taking the time to think through it on your own a bit. Which of these hypotheses (if any) would you consider possible to be proven wrong? That is, which do you feel are falsifiable?

1. Any two objects dropped from the same height will hit the ground at the same time.

This is definitely falsifiable. You just need to find two objects that hit the ground at measurably different times to prove it wrong. (Try dropping a piece of paper and a brick. Or better yet, a helium balloon and an anvil.)

2. No two snowflakes are exactly alike.

This is considered falsifiable because you just need to find two snowflakes that are exactly alike and you've proven the theory wrong. I do have an issue with the wording, though: Define *"exactly alike."* What does that really mean? At what level of detail do we examine them for sameness? If someone claims they've found two identical snowflakes, maybe they just *appear* to be the same width, but it turns out that my ruler isn't precise enough to measure the small difference between them.

Semantics aside, the thing to note here is, although you can prove this hypothesis false by finding just two identical snowflakes, *you can never prove that it is true.* Think about it: You can compare lots of pairs of snowflakes and they can all look radically different from each other, but does that prove that no two snowflakes have ever been identical, or will ever be? To prove the theory *correct*, I would have to compare every snowflake that ever was, is, or will be, against every other snowflake that was, is, or will be....everywhere in the universe.

This is worth repeating: **A valid scientific hypothesis is one that can be proven wrong**. You can never prove it to be true.

3. The government is hiding evidence of alien visitations.

How do I disprove that? Even if I searched every square inch of every government and military installation and found nothing, that doesn't disprove the theory. They could be constantly moving it before the examiners check a particular area. Or the evidence could be hidden elsewhere: In a hidden vault dug into the ocean floor, in a hidden vault in a secret base no which one knows about, etc.

It's funny how believers perk up when they hear that their claim can't be disproven. To them it's fantastic news, and it actually becomes their rallying cry: *You can't disprove it! You can't disprove it!*

I hate to burst your bubble, but if we can think of no experiment to disprove your claim then that's **bad news** for you, not good news. It forces your claim to be lumped unceremoniously into a massive pile of equally bizarre, non-falsifiable speculations, such as:

* The Egyptian god Horus was born of the virgin mother Isis.

* There is a separate plane of existence which we cannot detect.

* Magical fairies exist which become invisible when you try to detect them.

And of course...

4. **God created the universe.**

What experiment can I run which would prove that statement wrong? Obviously, the first step is to agree on an exact definition for what God is. But then, as I see it, I would have to be able to:

A) Go back in time to just *before* the universe was created.

B) Get *outside* of our universe so as to observe the creation event itself.

C) Run verification tests on whatever creator I detect there, to determine if it's indeed *the* god that we agreed upon in our definition (and not some other god like Zeus or Ra, or a super-powerful alien).

D) Assuming the verification tests from Step C prove the detected creator to be our agreed-upon God, I would then need to duplicate the conditions again *without God there* to determine if God indeed created the universe *or whether the universe's creation also happens <u>without Him</u>*.

193

Even if I did all that and the universe indeed came into being when God wasn't around, a Creationist could still say: "Just because the universe burst into creation when God wasn't there doesn't *prove* that He didn't do it. He just created it from wherever He was."

Sigh.

As you can see, there is no test that you can even *conceive* of to ever disprove the claim that God created the universe. Therefore, it's not a scientific statement.

You might forever dislike and distrust science, but it surely is the only tool we have to help us understand our world, and so our takeaway here is this: **If you can't think of a way for others to disprove your claim, then please label it as mere speculation when you announce it to the world.** This certainly applies to those who speculate that God (or an "intelligent designer") helped to guide evolution, which we look at in the next statement.

STATEMENT: The claim that an intelligent designer helped guide evolution is a valid scientific hypothesis.

AGREE:

You *agree* with that statement, even though we just discussed how a hypothesis needs to be falsifiable in order to be considered scientific (and therefore valid)? I can only assume you skimmed the first two sections of this chapter, so I'll pretend it's an honest mistake. And I'm happy to rephrase the main point. The important thing you missed is that *your claim needs to be vulnerable*. That is, someone who disagrees with you should be able to conduct an experiment which disproves your claim. The more he tries to disprove it, and the more he fails, the more valid your claim becomes.

We are forced to ask ourselves: **What experiment can we design to test your speculation that there's an intelligent designer who is guiding evolution?** You'll really want to think about this. I mean, you want to be sure that you're actually teaching *science* in a science class, don't you? So, play the role of skeptical scientist and think of an experiment where he can say, "Ha! Look at this result! This proves an intelligent designer was NOT involved in evolution!"

Can you think of one? Me neither. The best I can come up with is the following experiment:

1. I go back in time to just *before* life came into existence on Earth.

2. I round up *all* the intelligent designers, including the aliens that built the pyramids, the other aliens that built Stonehenge, the *other* aliens that erected the sculptures on Easter Island, plus the aliens that are abducting thousands of people daily, and of course the aliens that crashed their craft near Roswell. I'll also need to round up all the supernatural beings, including Poseidon, Thor, Zeus, God, Apollo, Ra, Yahweh, the Holy Ghost, Mithra, Horus (and many thousands more) and lock them in a room from which they can not escape or exert any influence on the external world.

3. Observe how (or if at all) life evolves without their influence.

The flaw in my protocol, of course, is that I can never be sure I've rounded up all the intelligent designers. Think about it. Imagine I indeed sequester all known aliens and supernatural beings and evolution *still* occurs. You can simply claim that there must be some *other* intelligent designer that I didn't find and isolate.

Another flaw in my experiment is that you can claim my room wasn't strong enough to negate your designer's ability to guide evolution. (Perhaps a signed waiver from each alien and deity would cover that aspect? Something to the effect that, "I, __*insert intelligent designer's name here*__, hereby attest that during this experiment I was sequestered and did not exert any influence on the course of evolution." Of course, even then, a believer would simply claim I forged the document.)

To some, the experiment described above might seem ridiculous and even condescending, but it's not. It is literally the closest we can get to creating an experiment to test *your* ridiculous and condescending "hypothesis" about an invisible, undetectable, magical designer.

Fortunately, the United States courts are in complete agreement on this. Time and again, our courts have ruled against Intelligent Design, declaring it a religious doctrine due to it being a non-falsifiable hypothesis.

DISAGREE:
Even if the courts shoot down every proposal to teach I.D., that's a small (though important) victory. The real battle is to win over the tightly closed minds of the misguided parents who force their irrational and wholly unsupported beliefs on their unwitting children. Because, sure, the state can force schools to teach evolution, but you can't force a Christian child to *believe* it. (I consider this to be the most important issue raised in this book, so we'll discuss it more in the final section.)

Personally, I'd love to take a "science" class at a Christian university where intelligent design is taught. Boy, would I have a lot of questions for the professor!

"Excuse me, professor. If our designer is so intelligent, why do we have dentists? That is, why are our teeth so poorly designed? Most people have overbites, crooked teeth, impacted molars and so on. And why are our eyes so terribly designed? Look how many people need glasses. And what an *idiotic* design to have the nerve fibers in our eyes *ahead* of the retina! It creates a blind spot where the nerve passes through the retina and out of the eye. Couldn't the designer have used the same design he came up with for the octopus? In octopus eyes, the nerve fibers do *not* block light or disrupt the retina. Any thoughts on these incredibly *unintelligent* designs?"

"Of course, my child" the Christian professor answers. "The intelligent designer works in mysterious ways."

Can you imagine how easy the tests would be in an "Intelligent Design" course?

1. *How did speciation amongst the birds in the Galapagos islands come to be?*

ANSWER: **The designer created them.**

2. *The modern opossum seems to have evolved from which cretaceous ancestor?*

ANSWER: **None. The designer created it.**

3. *How do homology and analogy differ in the way they bring about biological similarity?*

ANSWER: **They don't. The designer creates everything.**

I wonder if the "Intelligent Design" textbook gives equal time to all possible explanations for who or what the designer or designers might possibly be. After all, I.D. proponents aren't claiming that *their* god did it, are they? That would be an even more blatant violation of Church and State. They have to teach all supernatural possibilities. If I were outlining the I.D. textbook, the chapters would be as follows:

- **Ch. 1: Zeus is the Designer.**

- **Ch. 2: The aliens that built the pyramids are the Designers.**

- **Ch. 3: The aliens that abducted Whitley Strieber are the Designers.**

- **Ch 4: Xenu (from Scientology) is the Designer.**

- **Ch 5: Xenu's sub-commander of galactic sector 7 (hyper-quadrant 29e) is the Designer.**

- **Ch 6: The Zorgonauts that live inside Jupiter are the Designers.**

After a few hundred chapters "teaching" about possible supernatural designers, I'd then encourage the students to submit their own theories. In I.D., every theory is equally valid, right?

It's fun to joke about, but there's a deadly serious matter at hand. What's at stake is the mental health of school children. If you don't teach them to think critically (and if you withhold facts from them), you put their future education, their future career.....their future *period* at stake. Without critical thinking, their general ability to make informed decisions is severely at risk. And this extends to their *physical* health as well. In the next section, we'll see how.

STATEMENT: Childhood vaccinations are a direct cause of autism.

AGREE:
Not having children yet of my own, I can only imagine the agony that some parents must go through when deciding whether to vaccinate their children. It seems to be one of those "Damned if I do, damned if I don't" kinds of decisions. *If I give my child the shot, I'm told that I'm taking a risk that he'll develop autism. If I don't give him the shot, I'm obviously taking a risk he'll get mumps, measles, or rubella.* Since you agree with the statement that vaccinations cause autism, it's an agonizing decision that you, too, have made or will make.

There were two key words in the above paragraph. Did you catch them? "*I'm told.*" As in, "*If I give my child the shot, **I'm told** that I'm taking a risk that he'll develop autism.*"

The question we need to ask is: ***Who is doing the telling?*** Did you hear about a purported link between vaccines and autism from your friend, from your neighbor, or from a co-worker? Did you hear about this vaccine claim from Jenna McCarthy (a Playboy playmate) or from Jim Carrey (the actor)? Is your source of information some undisclosed writer on some random website? Why would you trust what *any* of those people have to say about something as important as your child's safety?

Think of it this way: If your child needed brain surgery, would you allow Jenna McCarthy or Jim Carrey to perform the operation? Would you call your neighbor and have her perform the operation? Would you hand the scalpel to your friend or your co-worker? Of course not. You would never put your child's life in the hands of someone who isn't qualified. And yet this is what you're doing when you listen to anyone who is not a board certified specialist in infectious diseases, nor an expert on vaccines, immunology, or virology.

That being said, in this book we've learned to ask: *What unexplained phenomenon exists to warrant such a belief?* That's a great question. In this case, there seems to be a very real increase—300% by some counts—in the rate of autism in the United States since the early 1990s. *What*, the believers ask, *if not vaccines could be causing this?*

There seem to be two obvious factors:

1. **We changed the way autism is diagnosed.** With a broader definition for what symptoms indicate autism, far more children suddenly received the diagnosis. To make a simple analogy, imagine if people are diagnosed as being "tall" if they are six feet or taller. If the definition of "tall" were suddenly changed to "anyone 5'10" or taller", wouldn't you expect a sharp increase in the diagnosis of tallness?

2. **There has been a large increase in funding for autism treatment programs and special education and programs.** That's fantastic, of course, but such programs give parents an incentive to seek a diagnosis if their child has any signs of developmental delay.

Even if you maintain that the two reasons above are not sufficient to explain the full extent of the increase in autism diagnoses, how do you make the bizarre leap to declare it must be vaccines? I'm not a specialist, but doesn't it seem more reasonable that something in the *environment* has changed which affects children who have some genetic predisposition to developmental disorders?

Of course, the claim that vaccines cause autism was first made by Dr. Andrew Wakefield in his 1998 study. *Lancet*, the very same (and otherwise respectable) British medical journal that prematurely published Dr. Wakefield's deeply flawed study on a sample of just 12 children, has now published a legitimate study of 498 children. The conclusion?

> Our analyses do not support a causal association between MMR vaccine and autism. If such an association occurs, it is so rare that it could not be identified in this large regional sample. (Taylor)

Before we rely on any scientific observation, though, it has to be reproduced. The above study has been performed at least twice more, with the same result: **The MMR vaccination does not cause autism.** (Offit)

By the way, it's not just your own child's health that's at stake. Immunizations are good for the overall public health, but if enough people decide to stop vaccinating their children, there develops a sizable pool for the disease to flourish in. In the U.S. there have been measles outbreaks in at least four states, located in such communities where the anti-vaccine mindset has taken hold. The official estimate from the CDC is that if we were to fully vaccinate all U.S. children born in a given year, it would save approximately 33,000 lives, and prevent 14 million infections, to say nothing of the $10 billion or so in medical costs saved, from their birth to adolescence. (Park)

Again we see the power of a single belief.

RESOURCES

For more information about vaccinations, please talk with a board certified specialist, and compare what you learn there with the information on a reputable site like the U.S. Center for Disease Control.

DISAGREE

It was the year 2000 or so when I first heard the claim that vaccinations are a cause of autism. A friend of mine in New York was my source of disinformation. He mentioned how his child had just turned six months old, and I asked if they'd gotten the immunizations out of the way yet.

"Dude," he said, employing the standard lexicon of immunologists, "are you frickin' kidding me? Vaccines are way dangerous."

His earnestness stuck with me, but since I didn't (and still don't) have kids, it didn't affect me much. He certainly didn't dissuade me from getting the flu shot every year. But his claim lingered in my mind, as do all extraordinary claims. Over the years I heard the anti-vaccination sentiment echoed here and there from others. There seems to be a movement, and a brief search online reveals its worldwide scope. Of course, Dr. Andrew Wakefield's "study" which started the hype in 1998 has now been declared as fraudulent by the publishers. That was the lone source of evidence for the "autism from vaccination" movement, but the fact that it has been debunked hasn't had much of an effect on the believers. (CNN)

I've lost touch with that friend of mine who first told me about immunizations, but I wonder if he's since reexamined his belief. One thing's for sure: I hope his children are okay.

STATEMENT: Certain meteorites discovered on Earth contain evidence of microscopic Martian fossils.

AGREE:

The meteorite named ALH 84001 has had quite a journey. It formed 4 billion years ago or so on Mars, and then about 15 million years ago it was blasted into space when an asteroid smashed into that planet. The small chunk of rock meandered our solar system for 13,000 lonely years before being caught by Earth's gravity and falling to the white plains of Antarctica. And there it sat until 1984 when a team of geologists spotted it and delivered it to NASA's Johnson Space Center, in Houston. Having endured twelve years of analysis, the meteorite was then escorted to NASA's press room where officials announced the claim of lead scientist David McKay that the meteorite contains evidence of microscopic Martian fossils.

I'll never forget that day back in August, 1996. President Clinton made an address to the nation about the discovery, and it made headlines in newspapers around the world. For years I held on to the New York Times article that stated, "Mars lived! Rock shows meteorite holds evidence of life on another world." For me it's one of those, "Do you remember when?"

kind of moments. But as eager as I'd been to believe their announcement, I remained skeptical. It's not like we're talking about worm-sized fossils. We're not even talking about paramecium fossils. These supposed fossils they're talking about are measured in nanometers. That is mind-bogglingly small. Just because the shapes they found *resemble* something doesn't mean they *are* that something. To paraphrase the astronomer Seth Shostak, "A cloud might resemble a horse. That doesn't actually make it a horse." (Moskowitz)

The fact is NASA spoke too soon and has since distanced itself from these types of claims. As of 2012, the consensus opinion of astrobiologists and others who are experts in the study of fossilized, microscopic lifeforms is that the evidence *isn't* sufficiently convincing. The magnetite crystals that are offered as evidence also form as a result of shock waves...the *same* magnitude shock waves that the rock was subjected to when it was blasted into space. (Garcia) It was also discovered that the meteorite had been contaminated by Earthly bacteria, which spread through cracks into the meteorite's interior. (Zimmer, 3)

The debate continues over this and other meteorites. But for now, if world class experts in the field are still debating it, we need to reserve judgment until a clear consensus is formed.

DISAGREE:
The claim which David McKay first announced back in 1996 was so monumental, and represented such an earth-shattering paradigm shift that other scientists were highly skeptical, and rightly so. If you announce to the world that you have evidence of extraterrestrial life, you'd better have no-doubt-about-it, incontrovertible proof. But nearly twenty years of research have come and gone, and the consensus remains that the evidence within the ALH 84001 meteorite (and other meteorites) is far from conclusive. (Zimmer, 1)

The topic of this chapter is the importance of falsifiable claims. So, is their claim falsifiable? Well, I have to ask: *Which claim?* Here's NASA's version, obviously aimed at the general public:

> The meteorite contains strong evidence that
> life may have existed on ancient Mars. (Jeffs)

Strong evidence? That's a bold statement. But is it falsifiable? No. First of all, define "strong evidence." Really, the word "strong" adds nothing to the hypothesis, because the whole thing is being qualified by the word "may." You could just as easily write, "The meteorite contains *incontrovertible* evidence that life *may* have existed on ancient Mars." We're still stuck with the word "may."

Heck, anyone can make the claim that life may have existed on ancient Mars without even *looking* at a Martian meteorite. Allow me: "Life may have existed on Mars." Prove me wrong. But no version of NASA's claim is falsifiable because **there's no experiment we can do** where the result might be, "Ha! Look! The meteorite doesn't contain evidence that life may have existed on Mars!" **We have no idea what Martian life might look like**, and therefore we can't possibly describe all forms of evidence for such unknowable life that are *not* in existence in the meteorite.

Meanwhile, in sharp contrast to NASA's bold version, here's the claim from David McKay and his team in their paper titled, *Origins of Magnetite Nanocrystals in Martian Meteorite ALH84001*:

> The majority of ALH84001 magnetites has
> an allochthonous origin and was added to
> the carbonate system from an outside source.
> This origin does not exclude the possibility
> that a fraction is consistent with formation
> by biogenic processes. (McKay)

"...does not *exclude* the possibility that a *fraction* is consistent with formation by biogenic processes?" Way to go out on a limb, guys. They've made an interesting statement, but there's nothing here for us to falsify. Again, ask yourself what experiment could be done where the result would be, "Hey, look! We've excluded the possibility that *any fraction whatsoever* of the magnetites (which we feel have an allochthonous origin) is consistent with formation by biogenic processes!"

Mind you, I find the debate fascinating, and the work that all these scientists are doing is a vital part of the long term investigation into the possibility of life on Mars. But the rational stance here is to side with the skeptical majority of astrobiologists. As of December, 1, 2012 there is no persuasive evidence that life ever existed on Mars.

STATEMENT: The universe appearing from nothing violates the first law of thermodynamics.

AGREE:
This is the argument used by Christian apologists who...

A) ...seem to think it's valid.

B) ...seem to think it proves the existence of a supernatural creator of the universe.

C) ...seem to think it further proves the existence of their exact god.

Before discussing their argument, I have to comment: Isn't the term *apologist* an odd (yet fitting) choice for a person who tries to explain and defend his belief in Christianity? "I *apologize* my beliefs are so bizarre. I'm so *sorry* that they make no sense to anyone. Please *forgive me* as I try to explain our three gods in one, our talking animals, and the singularly ludicrous claims about Noah's ark and the giant flood."

As for the topic at hand, give me one second while I climb to the top of my roof, turn on my loudspeaker, and shout out to the Christian apologists:

Who says the universe appeared from nothing?!!

This whole discussion relates to what's known as the "First Cause" argument, which we discuss in much more detail in Chapter 7. In this section, I only want to address the scientific aspect of it, as it relates to the 1st law of thermodynamics. So, allow me to repeat that key question: Where is the evidence proving that the universe appeared from nothing? *No one knows* what triggered the Big Bang, and certainly no one has any evidence that what preceded it was nothingness! My guess, for what it's worth, is that the creation of the universe was merely a transition from one state to another. Nothing was created in that instant of the Big Bang, only transformed.

But fine. Let's pretend the apologists are right. The universe came from nothing. Surely that would only violate the first law of thermodynamics if the energy that existed before the Big Bang was different than the energy afterwards. Well, it turns out that there are good reasons to think that the energy of the universe is indeed exactly zero! (Feuerbacher) The assumption being that the energy of the nothingness that supposedly existed before the Big Bang was zero, hence no violation of the 1st law.

Of course, what do I know? I'm not an astronomer. I have not one salient bit of insight to offer on the subject. Were I ever to offer an opinion about a cosmological idea, it should be given no more weight than, say, your dentist's views. But we can surely place value on what Gabriele Veneziano, a theoretical physicist at CERN has to say on the matter...

> [...at least two potentially testable theories plausibly hold that the universe--and therefore time--existed well before the big bang. If either scenario is right, the cosmos has always been in existence and, even if it recollapses one day, will never end. (Veneziano)

But fine, let's ignore all my arguments and concede that the topic statement is correct: The universe appearing from nothing indeed violates the first law of thermodynamics. All you can ever conclude from that is, "...**therefore it could not have come from nothing.**" That's it. End of story. No creator. No God in a white robe. No angels with trumpets. All you can say is, I guess the universe didn't come from nothing.

DISAGREE:
As I mentioned above, one glaring flaw in that argument is the unfounded assumption that the universe appeared from nothing. Another flaw is the assumption that the pre-Big Bang state and the post-Big bang state have differing energy values. I'd like to see proof of that, please.

What really irks me is the appalling hypocrisy of these Christian apologists. When they think something in science can be used to support one of their bizarre (and by definition non-scientific) speculations, they become boisterous advocates...

...of that *particular* theory, for that *particular* claim. Yet when that same exact scientific law disproves one of their other claims, the universal scientific law they were just relying on suddenly becomes invalid. Take for example Jesus and his magical production of bread. Can't we argue that...

> **STATEMENT**: Thousands of loaves of
> bread appearing from nothing violates the
> first law of thermodynamics.

Every time I bring this up to a Christian, I get this response: "God did that magically."

I see. Thanks for the clearing that up. Actually, for that lame explanation I could use another apology.

Climate change (a.k.a. global warming) is the most complex
issue I've ever researched. In no particular order, here are just a
few of the factors that need to be considered: Carbon dioxide
emissions, methane emissions, deforestation, sea level changes,
sea temperature changes, water vapor levels, changes in solar
output, past climate "proxies" (like tree rings), changes in the
earth's tilt and orbit, aerosols, cloud cover, melting of land ice,
precipitation patterns, and glacier retreat. How can I expect to
make sense of it all if scientists themselves are apparently still
weighing the data? And to whom do we turn? Certainly the
media is of little help. One moment they're sounding the
alarmism bell, the next moment it's "Climate Gate" and "The
Myth of Anthropogenic (human-caused) Global Warming."
Should we turn to the internet for information? If you do,
prepare to search through endless haystacks of bias to find
those few needles of truth.

After seesawing back and forth on what to believe about
climate change, I eventually settled on a position, but it's one
I'm certainly willing to amend as more evidence comes in. As
goes the consensus of experts, so goes my belief. So what is the
consensus? Well, it all depends on how you phrase the claim.
Since this is such a complex issue, let's break it into what seem
to be the three main points of contention.

STATEMENT #1:
There is little evidence that the earth is experiencing an
abnormal increase in temperatures.

<u>AGREE</u>:
What are your sources for information? I rely on NASA and
the National Oceanic and Atmospheric Administration, both of
which confirm that the average global surface temperature in
2011 was the ninth warmest since 1880. As it turns out, **nine of
the ten warmest years on record have occurred since the
year 2000.** (NASA)

The thing is, it's easy to make it *seem* like there's no warming
trend. You just have to pick a small time period and purposely
ignore the big picture. Scientists call this "going down the up
escalator." With this kind of cherry-picking of data points, you
could make it seem like there's no warming trend as we move
from spring to summer. For example, on April 11th of this year
the high temperature where I live was 82 degrees, yet on May
17th the high was only *79* degrees. I therefore conclude that
there is actually a slight *cooling* trend as we move from spring
towards summer. This is a dishonest evaluation of the data, and
is common among those who are trying to dispute the general
conclusion.

If you still believe that there's little evidence showing that the
earth is experiencing an abnormal increase in temperatures,
then please give your careful attention to the following sources.
Think of all that's at stake here. What if you're wrong?

<u>RESOURCES</u>

Climate Change And The Integrity of Science
Published in Science magazine, May 7, 2010.

NOAA: *State of the Climate, 2009*

A New Estimate of the Average Earth Surface Land
Temperature, Spanning 1753 to 2011, Berkeley Earth Project.

DISAGREE:
Of all the claims related to global climate change, this one is the least contested. Nevertheless, if you follow the global warming debate closely, you may have heard about climatologist Michael Mann's 'hockey stick' graph which has been used repeatedly to show that the earth is experiencing an abnormal increase in temperatures. Skeptics claim his graph is statistical nonsense, or that scientists haven't been able to replicate it. However, the National Center for Atmospheric Research did their own assessment of global temperatures over that same time frame, and came to essentially an identical conclusion, namely that the warming trend over the past thirty years is unprecedented for the past 500 years or so.(Cook, *What Evidence is There for the Hockey Stick?*) Of course, I'm preaching to the choir here, but if you find yourself in a debate with someone who is skeptical that we're in an ongoing warming trend, be sure to tell them: *If you look at all the data, it's clear that there is an unprecedented warming trend taking place.*

STATEMENT #2:

There is no scientific consensus that man-made greenhouse gases are the main cause of global warming.

AGREE:

Again I must ask: What is your source of information? I hope you don't quote the Oregon "Petition" (or any news source that quotes the Oregon Petition.) [The so-called petition is a list of 30,000 signatures of people who claim to have a B.S. degree or higher...*in any field*. It is directed to the U.S. government in the hopes of rejecting policies related to global warming.] In plain English, the Oregon "Petition" is an insult to the intelligence of any thinking person. The petition has fake names and celebrities on it, and anyone with a BS in virtually any field can sign it. Do you really care what Henry W. Apfelbach, an Orthopedic Surgeon has to say about climate change? The fact is, the vast majority of signers are completely unqualified in disciplines related to climate. (Grandia)

Meanwhile, a survey of all peer-reviewed papers published between 1993 and 2003 that dealt with global climate change showed that 75% of the papers agreed with the consensus position, while the remaining 25% made no comment either way, focusing on methods of climate analysis. (Oreskes)

214

The bottom line is that there's an amazing degree of consensus among the experts: **97% of active climate scientists agree that the current trend of global warming is primarily human caused.** Furthermore, the relative expertise of the dissenters (as measured by publications in peer reviewed journals and citations) was significantly lower. (Anderegg)

So far we've looked at consensus among individual scientists. But it's also worthwhile to look at the consensus of internationally recognized scientific societies. And once again, the consensus is nearly unanimous. Take, for example, what the U.S. National Academy of Science said in their 2010 report *Advancing the Science of Climate Change*:

> "Climate change is occurring, is caused
> largely by human activities, and poses
> significant risks for—and in many cases is
> already affecting—a broad range of human
> and natural systems." (NAS, Advancing
> Science)

That sentiment is echoed again and again in similar reports by similar scientific societies. (RealityDrop.Org)

And since the topic of this chapter is to ask whether claims are falsifiable, we should apply that to global warming, as well. If increased CO_2 is causing an increase in the amount of radiation that the atmosphere absorbs, we should be able to measure a decrease in outgoing radiation in certain wavelengths. Sure enough, when scientists took measurements of outgoing radiation with satellites, they indeed found this predicted decrease. (Harries)

If you're still on the fence about this particular statement, please take the time to research the all the citations above. It's important to reconsider your belief on this so that we can all be on the same page when it comes time to taking action against the possible consequences of global warming.

DISAGREE:

In this book I've used the term *consensus* numerous times, but you may have noticed I didn't define it. Of course, it has no strict definition, though it must be over 50% by any definition. But when it comes to climate change, the consensus is a very specific number: As I mentioned in the 'AGREE' section above, 97% of active climate scientists agree that global warming is occurring, and is caused mostly by human activity.

Those who are skeptical about global warming are fond of reminding us that correlation does not establish causation. Their point is this: Although levels of CO_2 have dramatically increased since the start of the industrial revolution, and although global temperatures have risen during that same time frame, that doesn't in any way prove that this rise in CO_2 *caused* the rise in temperatures. Although such reasoning is generally sound, if there's enough correlating data then statistical analysis can essentially prove causation, as was done between smoking and lung cancer. But in this case it's moot because **scientists aren't relying on correlation, they're relying on physics**. The fact is, it's been known for a long time that CO_2 is a greenhouse gas (meaning, it retains heat). (Cook, *Empirical Evidence That Humans Are Causing Global Warming)*. Once again I'm preaching to the choir, but it's also important to *reinforce* beliefs, especially on something so vitally important as maintaining the biosphere for future generations.

STATEMENT #3:
We need to take action to slow global warming to avoid serious consequences in the future.

AGREE:
I, too, agree with that statement. Of course, the trillion dollar question is *what actions*, specifically, should we be taking, both as individuals and as nations? Personally, I'm put off by the overly emotional propaganda that seems to accompany some people's interpretation of the scientific studies. Let's put aside the alarmist rhetoric and do some detailed cost-benefit analysis before implementing any drastic measures. In the meantime, no matter where one stands on this complex issue, wouldn't it be reasonable to develop our infrastructure to better harness our clean, renewable resources? Why aren't we developing massive solar power complexes in the deserts of Arizona and Nevada? Why aren't we developing equally massive windmill farms in the Great Plains? Even if the short-term costs are higher in harnessing and delivering solar and wind power, the long-term benefit of cleaner air and independence from foreign oil seems well worth it.

DISAGREE:

Even the most ardent global warming scientists can't say with 100% certainty that we're headed towards a bleak future. No matter how confident they are in their computer models and simulations, the farther ahead you try to look, the more variables creep in to make the predictions ever less certain. And besides, it's not like we fully understand all the factors that affect climate. They're still gathering data, still formulating theories. So in this, I agree with the global warming skeptics: There are so many uncertainties about global warming and its impacts. We don't know for sure what will happen.

Of course, science doesn't deal in certainty but rather probability. And the consensus of climate experts is that the probabilities are high enough to warrant taking action. Put aside all the hype about *possible* extreme weather events, *possible* negative affects from a more acidic ocean, *possible* droughts, *possible* fires, *possible* extinctions and diseases and food shortages. Some of those may well be true, but there doesn't seem to be consensus on them yet. But at a minimum, it's likely we'll have to cope with a significant rise in sea level which would effect hundreds of millions of people living in coastal cities and low-lying areas further inland. That alone is reason enough to take sensible steps of precaution. After all, *if you're willing to pay hundreds of dollars each month to insure your home in the unlikely event it gets destroyed, why not apply that same logic to your other home: Earth?*

Consider this very apropos analogy: There's only a small chance that a gigantic meteor capable of causing mass extinctions and threatening the human species will collide with the earth in the next hundred years. But there is obviously *some* chance. Are you against spending money to develop an asteroid detection system and action plan that could avert the disaster? The possibility of serious global climate changes are like the possibility of large asteroids hitting the earth. We don't know the size, we don't know the effect, and we can't even say for sure it will happen. It's just that there's so much at risk, we need to hedge our bets.

CRITICAL THINKING LESSONS
FROM CHAPTER 6

Here are the takeaways from Chapter 6:

* Science is the only tool we have to help us understand our world.

* If you rely on science to debunk ancient religions, then you need to be consistent and use it to disprove your *own* ancient religion.

* A strong scientific hypothesis is "vulnerable." That is, it can be falsified with a single experiment.

* Most hypotheses can not be proven right. But as we fail to disprove them, our confidence grows increasingly strong that they explain a truth about the world.

* Science is always open to debate. Its conclusions never absolute.

PARTING THOUGHT:
The Parallax of Antiquark Phenotypes

Over the summer of 2012 I read the book <u>Knocking On Heaven's Door</u> by particle physicist Lisa Randall. I was wanting to get some insight into what scientists are looking for with their multi-billion dollar machine—the largest machine in the world, in fact—called the Large Hadron Collider, near Geneva. The back cover assures us the book is written in an "easy to understand" style, but as I read it I realized that Lisa has an amazing gift for making the extraordinarily complex physics of quantum mechanics seem...*extraordinarily complex.* I barely understood a word of the book, at least not in any of the technical passages:

> *And whereas at lower energies, collisions involve primarily the three quarks that carry the proton charge, at higher energies virtual effects due to quantum mechanics create significant gluon and antiquark content.*
> (Knocking on Heaven's Door, pg 213)

Gotchya. Thanks for dumbin' it down for us, Lisa.

Actually, it's a great book, if a bit dense and jargon laden. Reading it, though, you get a glimpse into the incredible world of quantum physics, and a peek at the most fundamental particles of matter. It's a topic that draws me back to my childhood. Tell me, when you were a kid, didn't you ever cut a crumb in half, and then try to cut one of those halves in half again? Didn't you wonder where, or *if*, it ends? Is there a crumb so small it can't be cut in half? That's what particle physics is trying to answer. Is it complex? Yes. But it's also incredibly cool.

And that's my point. Sure, each branch of science has its own set of technical terms. Astronomy has its Cepheid variables and cosmic strings, parallax and perihelion. Not to be outdone, evolutionary biologists have alleles and phenotypes, epistasis and eukaryotic cells. But you don't need to know any of that stuff at first. There are all sorts of books and websites aimed at people with varying degrees of experience in science, all of which can help you discover the endlessly fascinating reality of the world we live in.

So, if you're new to science, where should you begin? Here are just a few resources to get you started:

RESOURCES

General Science:
One of my favorite books of all time is Bill Bryson's <u>A Short History of Nearly Everything</u> (2003). From superclusters to subatomic particles, Mr. Bryson writes about virtually every topic in science, adding in stories about the scientists themselves. It's necessarily a long book, but his writing shines with wit and clarity. It's really a must-read.

Astronomy:
Check out NASA's main website and pick a story that interests you. Personally, I love to follow the Mars rovers. http://www.nasa.gov/home/index.html

Evolution:
For a website which explains things with excellent graphics and straightforward text, I highly recommend Berkeley University's website called "Understanding Evolution": http://evolution.berkeley.edu/

For a book on the subject, one of my favorites is <u>The Blind Watchmaker: Why the Evidence of Evolution Reveals a Universe without Design</u> (1986) by Richard Dawkins. A better explanation for natural selection can hardly be found.

CHAPTER 7:
Let's Be Reasonable

INTRODUCTION:

If you're religious, you probably won't care much for this chapter. Chapter 6 may have taken a few shots at science, but the final chapter of this book aims squarely at religion. I honestly don't want to offend, and I certainly don't want to make enemies. Part of me wants to tell you, "***Skip it. Put the book down now and let's part as friends.***" If you do that, I certainly will understand. But I do hope you'll keep reading, objectively and impartially. There is just so much at stake here for all of us. Lives literally hang in the balance due to people's belief in religion.

$1/2(\cos(ax-bx)-\cos(ax+bx))-k/2(\cos(ax+bx)+\cos(ax-bx))=-1$

AGREE:

Are you sure you agree with that? Read it again!

DISAGREE:

So, you caught the mistake in the equation?

NO IDEA:

I put that equation in there to force people to say, "I have no idea." That's always bothered me about the most ardent believers: They "know" everything. It makes me want to scream: "Can you just *admit* when you don't know something?!" Why is it so hard for people to say, "You know what? That's not my area of expertise. I really have no idea how that works."

I'll tell you why. *People loathe uncertainty.* They will embrace the most bizarre and illogical things if it allows them to feel certain. But admitting your lack of knowledge is liberating. And it's honest. Just for fun, try saying the following out loud:

- "I have no idea what caused the Big Bang."

- "I don't know how life on Earth got started."

- "Evolution is a complex process dealing with things like chemistry, genetics, natural and sexual selection, and so on. To be honest, I don't understand it very well."

Isn't it both common sense and common decency to quietly defer to experts in a particular field instead of loudly contradicting them with your non-expert opinion? As the philosopher Ludwig Wittgenstein eloquently instructs, "Whereof one cannot speak, thereof one must be silent."

STATEMENT: There exists an all-powerful, supernatural, and undetectable being which magically created the universe. (Stated more simply: ***God exists.***)

AGREE:
This is as tough for me as it is for you. I have the unenviable— and likely insurmountable—task of trying to get you to reexamine your belief in God. To quote from the introduction to this book: *This guy thinks he can get me to turn my back on God? Good luck, buddy!*

I certainly don't look forward to the negativity some of you will send my way. (And before any of you send hate mail, please ask yourself if that's the Christian way to act. Does God really want or ***need*** you to say hateful things on His behalf?) And because believers love their God so passionately, I'm really not looking forward to hastening the end of your relationship with Him. So yeah, this is tough for me.

But it's tough for you, too. Rationalizing away the need for God is a bit like hearing how a magician did some amazing trick: "Oh, you mean that lady has a twin sister? So, during that burst of smoke, she actually hid in the base of that chair, and when her twin appears on the platform above the stage, we assume it's her. I get it now." It feels great to finally know the truth, but it's also a big letdown. It was *fun* believing, wasn't it? It was comforting to think there is some magic power out there. And it sure would be great to think that there's more to come after this life.

The truth is, growing up is always sad. If you've ever watched a child come to the realization that Santa Claus is actually something adults made up to control kids ("Santa is watching, so be good!"), then you've witnessed that bittersweet moment as reason overcomes faith. It *is* sad. But it's a necessary part of growing up.

Alright, so much for the buildup. We're going to discuss the existence of God now, so if the topic upsets you, this would be the time to skip to the next section.

Still here? Okay, so...You believe there exists an all-powerful, supernatural, and undetectable being which magically created the universe. But remember the question we need to ask about any particular claim: **What reason is there to believe that?** Which unexplained phenomenon is your claim shedding light on? To me, the only possible issue is:

What caused the Big Bang?

I don't know. At the moment prior to its expansion, our universe was a spectacularly energetic point; a *singularity*, to use the physics term. And that's what black holes are, too: Black holes are single points in space with an unfathomable amount of matter packed inside them. Perhaps our universe is the result of an expanding black hole in some other, larger universe.

Along with whatever challenges such a theory poses to physics, it still leaves us with the question: Where did that other universe come from? Ultimately, there are only two possibilities:

Either matter and energy have always existed, or they appeared from nothingness. Both of those concepts are really hard to conceive of, but that doesn't in any way suggest a supernatural origin, does it?

Allow me to plug my ears as the believers shout, "That proves there's a god, because everything has a cause! You can't get something from nothing!" Meanwhile, the skeptics in the room are firing back with, "Then where did God from? If God can always exist, or pop into existence from nothing, then so can energy!"

While those two go at it, let's you and I step into the hall and talk quietly...

I'm sure you've read other books on this "First Cause" issue, and have read plenty of online forums and discussions about it. You know how both sides of this old argument go. I don't have much to add, except this: It's impossible to have a meaningful discussion about something if the thing in question (in this case *God*) has no definition. So, I'm hoping you agree on this one: *"God is an all-powerful, supernatural, and forever undetectable being which magically created the universe."* By definition that's not something we can ever test, but it'll have to do for now. And regardless, we need to remember the main point here: What reason is there to believe that a supernatural creator exists?

The only unexplained phenomenon is the Big Bang itself, and that's at best a stalemate for you. But since my explanation (matter and energy **have always existed**) is far simpler than yours (a super-natural being of immense complexity, which has the ability to exist outside the universe and is able to manipulate matter and create the expanding universe *has always existed)*, then mine is the one we should accept. With my explanation, you only need to accept one incredible stipulation. With yours, we need to accept *numerous stipulations which are even more incredible.*

I've stated that there's only one unexplained phenomenon (the Big Bang) that could possibly warrant the belief in a supernatural creator. Many will take issue with that, claiming the following as another unexplained phenomenon:

Why does it seem that the laws of physics are fine-tuned to result in a universe that's favorable to life?

To me it's a non-issue. (And again, I really have to stress: I am not a scientist, but instead merely a rational thinker.) *The supposed "fine tuning" argument assumes that life is the purpose of the universe.* What an appallingly self-important idea that is! In any case, it's obvious that life is not the purpose of the universe, otherwise it would be ubiquitous.

The fact is, if your Creator did the fine tuning, he was terrible at it. **With the exception of Earth, the universe is incredibly hostile to the very life it's supposedly been "fine tuned" for!** If Christians disagree with that, and instead feel the entire universe has been magically tuned to welcome life everywhere, then please, by all means, go live on Venus where the surface temperature is 800° F, and where there's no oxygen or liquid water. If Venus isn't to your liking, then please consider the wonderfully fine-tuned atmosphere of Saturn, where you can go live in -400° F ammonia ice clouds. "Fine tuning" really is a ridiculous thing to claim and should be rephrased:

> "The universe has been *terribly tuned* so as to just *barely* support life on one immeasurably small speck of rock for a small portion of its limited existence."

The fine-tuning argument also assumes that there are no other possible values for the fundamental physical forces (like gravity, or the strong nuclear force) which would result in life. Really? Can you prove that? And who says life has to be carbon based, anyway?

Furthermore, if you allow for multiple universes with varying parameters (the multiverse theory is supported by many astrophysicists and apparently has a chance of being observed), then one of these universes is bound to have the parameters that result in a place somewhat hospitable to life.

To summarize: There's no reason to postulate an eternal creator. The one true mystery, the Big Bang, is explained more satisfactorily with the simple explanation (matter has always existed) than your extremely complicated and far more fantastic explanation.

I've had this kind of discussion with many Christians, and at this point of the conversation, many fire back with something very interesting. In what seems like a radical change of topic, they ask, "You love your mom, right? You love your wife?"

"Yes. So?" I ask.

"I bet you can't prove it," says the Christian. "So does that mean *love* isn't real?"

That's an excellent point, but ironically it proves *my* stance, not yours. Love is indeed simply a label we give to a large set of neurochemically induced feelings. Love is located and constructed entirely within the brain. Take away the brain, and love no longer exists. The same is true of the feeling called "God." God is a label certain people apply to a large set of neurochemically induced feelings. God is located and constructed entirely within the brain. Take away the brain, and God no longer exists.

That's my final answer: **God exists only in your brain. He is only as "real" as love or anger or jealousy.**

As always, the goal of this book is to get people to reexamine their core beliefs. I'm sure I didn't persuade you to question your belief in God any more successfully than all those other books you've read, and all those other discussions you've been following on the net. But that's okay. Skepticism is a plant that takes root in doubt, gets nourishment from logic, and needs the bright sunlight of reason to nurture its growth. I just hope I've been able to plant that first seed.

DISAGREE:

The origin of the Big Bang is the only unexplained phenomenon that believers in a god can rely on to support their claim. Yes, the start of the universe *is* a mystery and it might always remain one. But the postulation of an all-powerful, supernatural being as an explanation only creates tougher questions than the one it purports to solve.

You and I know that. It's self-evident. But it's sad the way Christian apologists try to doubletalk their way out of the vault of logic they've been trapped in. It hurts my brain trying to make sense of their long, rambling "proofs."

Here. *You* have a go at it. Below is the Christian explanation for the cause of the universe:

The Christian Apologist Explanation
If the universe has a beginning then there must be something outside it that brought it into existence. The cause of the universe must be a supernatural being that exists outside of time and yet is able to manipulate and control matter and energy using its magical, supernatural powers. How else could a temporal effect arise from an eternal cause? If the cause were simply a mechanically operating set of conditions existing from eternity, then why would not the effect also exist from eternity? Obviously this proves the Christian God created the universe.

Matter or energy have always existed.

Again, notice how extraordinarily complicated the Christian answer is. As always, that is the indication that it's a human fabrication and not a factual answer.

This "First Cause" debate has been raging for centuries, but if I may, I'd like to finally put it to rest. Let's look at a typical analogy that Christians use to establish the premise of their "First Cause" argument. *Imagine a sandcastle,* they tell us. *It didn't always exist. Obviously it had a creator.* From that they deduce:

Everything that comes into existence must have a cause.

That's fundamentally erroneous thinking right there. Why? Because I submit that ***nothing comes into existence***. Matter simply gets reshaped. To refer back to their sandcastle line of reasoning: ***The kid didn't create the sand, did he?*** Nothing was created. The pre-existing sand was just shoveled around. And therein lies the truth about the origin of our universe. The simplest answer is that matter and energy have always existed. Whatever triggered the Big Bang, it was merely a transition of energy, not the creation of it.

When I was a kid, the theory back then was that there was enough matter in the universe to slow its expansion, and in time actually *reverse* it. Such an event was labeled The Big Crunch, and that seemed like such a majestic solution to the mystery of the universe's existence: A never-ending series of Big Bangs and Big Crunches. It was the belief I held for a long time....until the scientific consensus changed, telling us that there apparently is not nearly enough matter in the universe to slow its expansion. There will likely never be such a Big Crunch. And thus, out the window went that graceful idea. Unlike believers, I'm willing to change any belief. It all comes down to evidence.

Although it's tiresome, what I do like about this "First Cause" argument is that it gives me hope. It shows that at least **some** Christians accept that the universe indeed started with the Big Bang 13.6 billion years ago. I'm inclined to let them keep their fantasy that a magical supernatural being created the universe if they'll allow the rest of us to teach astronomy and evolution unimpeded in public schools.

* * * *

You've been reading this section because you agree there's no reason to postulate the existence of an all-powerful, supernatural being which magically created the universe. Either my skills of persuasion in the 'AGREE' section were so powerful that I convinced a believer to change his mind, or you, too, are a rational thinker. In any case, notice how I don't use the term *atheist*. We'll discuss this "word" in the next section.

STATEMENT: If someone doesn't believe in God, they are an atheist.

AGREE:
I have three words for you: Prove God exists.

Others have made this point but it is worth repeating because its importance can't be overstated: ***There is no such thing as atheism.***

"Sure there is!" you might say. "Atheism is the non-belief in a God."

Sorry, but that helps not at all. You need to prove God's existence first, or neither word has meaning. Think of it this way: Do you know what a snark is? Do you believe in snarks? No? Then you're an asnarkist. God is as imaginary as a snark. Thus, the concept of atheism is as meaningless as asnarkism.

Put yet another way: **How dare you label me as a non-believer of something when you yourself can't produce evidence of the very thing you accuse me of not believing?**

Imagine if someone were to point at you angrily, yelling, "This guy doesn't believe in Krag, the nine-legged polka-dotted bird that lives on planet Zorgon!!! He's an a-kragist!!!!"

"Atheist" is a condescending and insulting label created by the superstitious denizens of the dark-ages. You should never label yourself an atheist or even recognize the word. And if a believer labels you as such, stand firm and deny it. Tell him, "*You* are the one making the claim for which you've presented no evidence. *You* are the one in need of a label, so let's be accurate about things. We'll call you '**a believer in a supernatural being that magically created the universe.**' Whereas myself, I'm simply a rational person waiting to see evidence of your fantastic claim, along with all sorts of other fantastic claims."

I'd press him further. "How does it feel being labeled?" And then I'd introduce him to my rational friend, "Hey Bill, I'd like you to meet a Christian friend of mine. He's a believer in a supernatural being that magically created the universe.'"

"Wow, there's people still like that in the world?"

I'm passionate about this point. If you're so fond of labeling people based on their non-belief in non-existent things, you need to be consistent and introduce *yourself* that way, too. "Hi, my name is John. I'm an...

...a-krishna-ist
...a-ghostist
...a-leprechaunist
...a-Baba Yaga-ist
...a-bigfootist.
...a-apolloist (Greek god of the sun)
...a-Loch Ness monsterist
...a-horusist (Egyptian god of the sky)
...a-alien-bodies-in-Area-51-ist
...a-poseidonist
...a-xenu-ist

...a-Santa-Clausist
...a-unicornist
...a-mithraist (the Zoroastrian god of truth)
...a-goblinist
...a-hadesist (Greek god of the underworld)
...a-fairyist
...a-seven-legged-elephantist
...a-all-powerful-being-that's-more-powerful-than-my-own-God-ist
...a-zeusist
...a-zarathustra-ist
...a-thorist
...a-ra-ist (the Egyptian sun god.)
...a-tooth-fairyist
...a-flying-giraffe-ist."

Those, plus an infinite amount of equally non-existent "ists" which you (I assume) do not believe in.

Either we need to label all people this way, listing all of the non-existent things they do not believe in, or we can save a lot of time and put the labels on those making the positive claims. If you agree with that, please read the section below. If you still insist on labeling people, then at least finish labeling yourself before moving on.

DISAGREE:
It's refreshing to talk with a rational thinker like yourself. If you skipped the above discussion, allow me to reiterate the main point: Believers shouldn't get away with using synonyms which hide the oddity of their belief. Force them to admit what they believe in detail. Don't let them slip by with, "I believe in **God**." Make them spell it out: "I believe in **an all-powerful, supernatural being which magically created the universe.**"

And this is crucial: ***Do not label yourself as an atheist*** because it tacitly supports the concept of a god. Remember, they're the ones making the unproven claim, so they're the ones in need of a label. In your interactions with Christians, this crucial part of your conversation should go as follows:

<u>Christian</u>: "What're you, an atheist?"

<u>Rational Thinker</u>: "Nope. But I take it you believe in a supernatural being that magically created the universe?"

<u>Christian</u>: "Uhh...well, yeah."

<u>Rational Thinker</u>: "Nice to meet you. As for me, I'm a rational person waiting to see evidence of your incredible claim."

If you can rattle that off in a conversation with your religious friends, it will give them pause. So be sure to memorize your lines. Rational people everywhere are counting on you.

STATEMENT: God is *not* allowed to change his mind.

For example: Since the prophecies predict Jesus' return in a particular way, with lightning and angels and such, it *has* to be that way. **God can NOT change his mind**.

<u>**AGREE**</u>:
Instead of viewing God as the director of this great stage play we call life, most Christians seem to consider Him merely an actor, forced to say His lines and act out His role exactly as written in the Bible. But a god who is not allowed to change his mind doesn't sound all that powerful to me. It sounds like your god is an unthinking robot, doomed to carry out an ancient program; a slave who's unable to make his own choices. He *must* do what is written in the scriptures. *He is powerless to think. Powerless to reconsider. Powerless to change His mind.*

Be honest: Does that sound like an all-powerful deity to you? And frankly: *How dare you to presume to know the mind of God!* If He wants to change His mind—prophecies or no—surely He can.

My guess is, Christians will rationalize this by saying, "Well, God *could* technically change his mind, but He *promised* us that Jesus will return in a certain way. And a loving God will definitely keep his promise." But the fact remains: Since your god *has* to keep his promise, then he is on a leash. He doesn't even have the choice to re-think a 2,000 year old promise. You might have "free will," but your own god does not.

You have to admit that perhaps God is having second thoughts about the Second Coming. It's been 2,000 years and counting, and still no Rapture. Maybe He has seen how we've progressed via the free will He gave us, and He really is re-thinking the whole Armageddon thing. This is an *almighty* god we're talking about. Isn't it at least *possible?*

You agree, now? It's possible? Great! Then please read the section below...

DISAGREE:
So, we agree that God *can* change his mind if He feels like it. And if He so chooses, He's allowed to break some ancient promise He made to a bronze-age tribe of sheep herders. I apologize for the sarcastic tone, but you wouldn't believe how many Christians I've spoken with who will not allow God to change His mind about any prophecy...*even though He's the one who wrote it!* He's God, for crying out loud. Surely He can do whatever He wants, and doesn't need to explain His reasons to you!

So, we're all in agreement, yes? **God can change His mind.** In which case, please go to the next statement...

STATEMENT: **Given that God is allowed to think for Himself and change His mind,** it is quite possible that Jesus may never return at all. Or, if Jesus does return, it will be in a radically different manner than as described in the Bible.

DISAGREE:

Darn! Just when I thought we were making some progress. Now it seems you're back to viewing God as a slave to prophecy, with no ability to think for Himself? In which case, please re-read the previous discussion where you granted God the freedom to change his mind.

I'm sorry to send you back there, but your thinking is caught in a loop. You'll need to reconsider this important point about God's freedom before continuing through the book.

AGREE:

"Sure," replies my Christian reader. "I guess I can go along with that. God can change his mind. But I believe He won't. I believe Jesus will return in fulfillment of the Scriptures."

Here's the key point: I'm okay with that, as long as you mean that you "feel" Jesus will return. (Remember our discussion on the two meanings of the word "believe.") You are not stating it as fact. You accept Jesus might not ever come back. It's simply your opinion (and I suspect also your hope) that He will.

It seems like a minor concession, but it's actually a major victory. By admitting that God is allowed to change his own Biblical prophecies without consulting us humans first, you've taken that important first step in developing your rational thinking skills. Do you dare take it a step further? If so, then I'll see you on the next page...

:
Imagine a crowd has gathered around a man standing on the beach. You join them and marvel as the man, dressed in a white robe, walks along the surface of the water. Returning to the beach, he raises his hands and parts the sea. You can see a twenty foot wide strip of the sea floor extending to the horizon between the two undulating walls of sea water. The man in the robe commands the parted sea to return to normal, then scoops up some water into a jug. From that jug he pours out *wine* into a glass.

STATEMENT: This person is God.

AGREE:
For me, it all depends on the definition of God. If **God** is defined as: "*Any being capable of walking on water, parting a large body of water, and transforming water into wine, (or any being that has access to technology which allows him to do the aforementioned),*"...then YES, that man on the beach is God, or at least *a* god.

But what I'm wondering is this: Even after such a display of power, how do we know he's actually *God*, and not just an incredibly powerful alien? I'm not saying this in any mocking way. I'm dead serious. Imagine the events of Rapture, where we see a being apparently controlling lightning, surrounded by humanoid beings with wings (what I'd label as angels). Isn't it important for Christians to ask, "How do we know it's *God*, and not actually a super-powerful alien being that's just pretending to be God?"

Phrased another way: **What test do you give to differentiate between Jesus and a super-powerful alien?**

I am not making fun of anyone's beliefs here. It's a very serious question, and one Christians need to ask themselves. After all, if some very powerful alien were to come to Earth and pretend to fulfill the prophecies, and you begin to worship this imposter, there could hardly be a worse sin! *You'd be taking a false god*.

I imagine a typical Christian response would be, "That's ridiculous! Why would an alien being do that? When Christ returns, we'll all know it. We'll all *feel* it. Now stop asking stupid questions!"

But such a response doesn't answer the question. You might feel in your heart that it's Jesus floating around up there, surrounded by lightning, and angels with trumpets, but that's all it will be: A feeling. You won't know it's really Him, because *there is no test you can give to differentiate between God and a super-powerful alien*. None. No matter what you ask of God to prove Himself, you will never know if it's really Him, or just an incredibly powerful alien pretending to be God.

This whole time, in fact, you may well have been praying to a false god. I submit that the "God" of the Old Testament could very well have been a powerful alien being. To me, that would make much more sense, especially considering His vindictive, cruel, and often immature behavior. Plagues, genocide, mass drownings, etc. That sounds like some powerful but petulant alien being *pretending* to be God, doesn't it?

How would you feel if the *real* God came to Earth tomorrow, picked up the Bible, and said, "I didn't write any of this! All that stuff—the flood, Moses and the Red Sea, people turning into pillars of salt, Jesus waking from the dead—that was all orchestrated by an alien named Bozhe, from a planet about 80 light years from here. I can't believe you people fell for this! Don't you have any judgment?"

He flips open to the Ten Commandments, "'*Thou shalt not worship false idols?*' Why would I care who you worship? And this one, '*Though shalt not kill.*' Do you really need to have that written down as a law? You need a *book* to tell you that killing is wrong? Are you serious?"

(It's a great release for me to write from God's perspective, so allow me to go just a little further.)

...God slams the Bible shut. "I've never written *any* so-called 'Holy' book. But if I did, it wouldn't be filled with genocide, baby murder, and the endorsement of slavery to name just a few of the disgusting things described in this book! I would've written a chart of the elements I created, explained the grand unified theory of physics, and given proof of the multiverse hypothesis. I would've shown how to cure cancer and AIDS and a host of other debilitating diseases. And I would've pleaded that everyone show tolerance towards each other. Wouldn't you have done the same?"

DISAGREE:
As I mentioned in the 'AGREE' section, whether our miracle worker on the beach is God or not depends on the definition of God. Of course, usually the definition of God includes the character trait of being *all-powerful*, in which case our water-walking, sea-parting man hasn't proven himself yet. But can he ever? How do you prove that a being is all-powerful?

I know this seems like a ridiculous thing to discuss but I actually think it's important. These believers are the ones who keep droning on about God and the Second Coming, so why can't we ask some probing questions about it? You want to make a ridiculous claim? You want to make a lot of noise about your almighty God? Then you have to expect to be taken to task for that. Let's take a nice, close look...

Pretend our apparent beach-God agrees to be tested. We escort him to our lab and set out to disprove the hypothesis, "This supernatural being is all-powerful." It's a good hypothesis because it *is* falsifiable. We just need to find one thing he can't do, and the hypothesis has been proven wrong. If I were one of the lab techs running the experiment, I'd start with the basics:

Task #1: "Move Mt. Everest to the Mohave desert, but don't hurt anyone in the process."

Task #2: "Please cure every man, woman and child in the world who is currently suffering from a fatal disease."

Task #3: "Please restore the limb or limbs of every amputee in the world."

Unfortunately, no matter how many of the tasks he accomplishes, we can never prove the hypothesis that he is God because we can never request all possible tasks. Who knows if the next task we ask of him will be too difficult?

A more serious issue though, is this: ***We'd never know if it was really the supernatural being that accomplished each task.***

* Maybe this guy has a spaceship, and there is powerful equipment on board that his cohorts are using to accomplish each task.

* Maybe a second, more powerful super-being is accomplishing each task and letting our beach-God simply *think* he's the one who's doing it.

The end-of-discussion, bottom line, fact-of-the-matter is this:

> **No being can *ever* be declared to be all-powerful because it would require an infinite amount of simultaneous demonstrations.**

That's right. He'd have to be able to accomplish literally everything—every single possible thing that can possibly be imagined—*all at the exact same moment*. If he can't, he's not all-powerful.

(Can we please drop this ridiculous notion, now, of all-powerful beings?)

STATEMENT: Before committing to their particular religion, most religious people carefully weighed all their possible choices.

For example: If you are Christian, you chose that religion only *after* doing extensive research into Buddhism, Catholicism, Confucianism, Hinduism, Islam, Judaism, Mormonism, Scientology, Sikhism, Wicca, and so on. You read all their texts, then pondered the logic of, and evidence for, their core beliefs. After meeting with a representative from each religion, and attempting to establish a relationship with each god, you finally concluded that Christianity was the correct religion.

DISAGREE:
I'm going to assume universal disagreement on this one. Phrased the other way: Surely we all agree that **if you are religious, then you did not weigh all your choices before choosing your religion.** It's doubtful you even researched the other branches and denominations of your own particular religion. For example, if you're a Methodist Christian, did you honestly consider the 7th Day Adventist denomination? Or Southern Baptist? Or Pentecostalism? Or Christian Science? Or the dozens of other branches?

I think it's fair to say that most people are simply a member of the religion that they were raised in. In some ways, that's as obvious a statement as saying you speak the same language that you were raised in. We don't choose our native language and we don't choose our religion. Sure, you might've been raised Catholic and then switched to a form of fundamental Christianity, but that's merely jumping between two branches of the same tree. It's not like you made the leap over to Zoroastrianism, (which was one of the largest religions in the world, by the way. Older than Christianity, too. And yet doubtful you even considered it.)

This section is about choice, and in short we can say: If you're religious, your *parents* chose your religion for you. But that's a bit strange, isn't it? Why should parents have the right to force their children to believe the same thing they believe? Did your parents force you to love the same music they do? Did they force you to have the same favorite authors, favorite movies, favorite sports? Did they decide who you should be friends with? *Did they tell you who you must love?*

Shouldn't we all have the right to choose—especially something so personal as a religion—without it being forced on us when we're the most vulnerable and most impressionable?

Imagine if all children were raised without any religious indoctrination of any kind. That means no taking them to church, no Bible readings, no prayers, and so on. Then, on their 18th birthday, they are taken to a huge building called the "Religions of the World Exhibit." In each room, a religion is represented which has the texts of that religion, a list of their beliefs, and a spokesperson to whom questions can be asked. There'd be a Judaism room, a Shinto room, a Unitarianism room, a Cherokee room, a Sunni Islam room, a Greek Orthodox room, a Scientology room, and on and on. Every single religion, past and present, is represented. (There'd also be a Rationalism room, where they can learn that there's no reason to choose any religion at all.) Each 18 yr old would be free to wander the building, researching which, if any, religions he might consider joining. I mean.....***shouldn't it be that way?***

Of course, one can hear religious parents the world over saying, "No! It shouldn't be that way! I can't have my child worshiping the wrong god! I could hardly think of a worse fate!"

I understand that reaction. If you're deeply religious, it probably sickens you to imagine your child not knowing and loving your particular god. But despite that heretical idea I just proposed (i.e. allowing people to choose their own religion independently, upon reaching adulthood), you're still reading this book. That means you're still willing to listen. So let's continue with our thought experiment. **What would happen if children were raised without religion until their 18th birthday, and then were allowed to research all possible religions, without guidance or bias?**

Imagine yourself in that situation: Pretend you were raised completely without religion, so much so that you didn't even know what religion your parents or siblings adhered to, or ***if they even followed one***. On your 18th birthday you enter the "Religions of the World Exhibit," and as you walk the halls you discover that there are thousands of religions and gods to choose from. What would happen? Which religion—if any— do you think you'd choose in that situation? Remember, you have no idea if anyone believes any of this anymore. You're on your own to decide.

It's worth a minute to really ponder that.

Given that I was raised Catholic and yet still became rational at an early age, I feel confident I'd follow the same path. I'd walk out of the "Religions of the World Exhibit" better informed about the creeds of some religions, but unconvinced by their myths. And though the representatives would speak with passion, none of them would be able to offer any sufficient proof of their claims. Ultimately, I'd once again conclude that there's just no ***reason*** to believe any of it.

But what about you? Assuming you're religious, do you really think that out of all those hundreds and hundreds of religions you stroll past in this "Religions of the World Exhibit," that you'd settle on the same one you have now? In reality, such an exhibit would be so huge, you likely wouldn't even **happen upon** your current religion. Your legs would tire long before that, to say nothing of your mind.

I suspect most people would hold off on their decision until they consulted with their friends or family. But those who are inclined to believe in something would probably pick one of the simpler ones. It's just a guess, but I think the less demands and claims that a religion makes, the easier it'd be to accept if approaching it cold, in your adulthood. That is, the closer a religion gets to saying simply, "Our only belief is that an unknowable being created the universe," then the more likely it is to sway critically thinking adults. But start piling on magical claims, like...

* We believe in a magic flying horse.

* We believe in magic glasses that are used to read magic golden plates.

* We believe in an immortal being who nevertheless was killed and magically came back to life.

...and you will raise the eyebrows of all but the most gullible adults.

This book, of course, is about questioning one's beliefs. And while I'm not likely to get you to reexamine any of your core beliefs, perhaps you'll reconsider forcing your beliefs onto your children. That alone would be a huge victory for everyone.

STATEMENT: Parents should be allowed to decide everything that their children are taught.

AGREE:
Did you know that in 2012, schools in Georgia and Louisiana began teaching that the Loch Ness monster is real, and thereby "proves" humans and dinosaurs co-existed? (Sherriff)

Does that not sicken you? Taliban schoolkids will now surge ahead from dead last in the world in science and math, leaving that spot to kids in Georgia. (Not that the Christian parents care.) Anyway, the essence of my question is this:

Should parents be allowed to teach their children anything they want based on the assumption that they *own* their children? Or, as I would argue, should parents respect their child as an independent human being who might appreciate hearing established facts about the world they live in?

Am I wrong on this? Maybe parents really *do* own their kids, and can therefore tell them any story they want as an "explanation" for the real world, such as:

* The earth is a group of large, flat islands floating in a huge bowl of water.

* The sun is a ball of fire that the sun fairy pulls across the sky each day.

* The sky is a big dome.

* Stars are just little holes in the sky dome where light shines through.

* Rainbows are painted by leprechauns as a map to find their buried gold.

At this, the Christian rolls his eyes. "Don't be ridiculous! Rainbows are a promise from God that He will never again kill every single man, woman, child and animal on Earth in a magical flood."

* Everything in the world magically came into existence 6,000 years ago.

* The dinosaur bones we find buried deep in the ground were put there by the magical sky fairy who wants to fool us into thinking the earth is much, much older than 6,000 years old.

"Damn right!" declares the Christian parent. "He's *my* kid, he'll learn what I decide!"

As a final plea to reason, didn't we decide in the statement regarding faith in Chapter 1:

> Faith is all a person needs to know the truth...
>
> ....as long as the issue doesn't relate directly to my health or safety or finances, *or to my child's understanding of math, basic physics, basic chemistry, and history.*

If you agree with that, then you have to agree that parents should **not** be allowed to decide what their children are taught. Certainly not regarding any provable facts about the world. If that's the case, then please skip down to the final 'DISAGREE' section below.

However, if you still feel that your children are yours to teach as you please, then continue reading this section. But the gloves are coming off, now. I've tried to restrain myself as much as possible throughout this book. I've tried to keep things civil and rein in the sarcasm. But we've come to the end. If none of my other arguments have made you reconsider your fundamental beliefs, then I've got nothing else to lose. I might as well speak from the heart, and tell you how I really feel.

Clearly, America is torn. We've been divided before, of course, but now the issue isn't slavery, it's religion. To resolve this, *I advocate dividing the United States into two countries.* The Rational States Of America, and the Religious States of America. Everyone would surely be happier. In the Rational States, we'll teach kids actual facts about the real world. The laws of chemistry, the facts of evolution, etc. Meanwhile in the Religious States, they can teach about the Loch Ness monster, talking snakes, and magical floods.

One thing, though. We'll be wanting our instruments back. You guys—er, *y'all*—will have to live without anything that the rational thinkers of science developed. That means you'll be living without electricity, without medicine, without x-ray machines or CAT scans, without cars, cellphones, computers, the internet. Think *Amish* but minus even the glasses or contact lenses. Scientists are also the ones who invented guns and rockets and tanks and drones, so we'll be taking all those with us as well. You might worry about being able to farm efficiently enough to feed yourselves, (though I don't see why: a loving God would never let anyone die of hunger, would He?) But at least you won't have to worry about being cold. The libraries in the South certainly have plenty of science books y'all can burn.

DISAGREE:
I had this discussion with a Christian friend of mine, via email. His main point was: We can't really prove anything with 100% certainty, so whose "truth" should we be teaching our kids?

Great question. Here's the answer:

The truth that should be taught is the one for which there is evidence. So, if parents want to teach their kids that the earth is flat, I would argue that that's irresponsible. We can prove the earth is round. And if parents want to teach their children that the earth is 6000 years old then that's also irresponsible, and frankly selfish. We can prove the earth is 4 1/2 billion years old, and the universe is 13.6 billion years old. And we can prove with literal and figurative *mountains* of evidence that evolution is real and ongoing. It's not in dispute by a single peer-reviewed scientist. By putting "truth" in quotes are you implying that there are no consensus views on what is factual?

If you skipped the 'AGREE' section above, here's what I'm furious about. In 2012, schools in Georgia and Louisiana began teaching that the Loch Ness monster is real, and thereby "proves" humans and dinosaurs co-existed. Here are two articles on the topic:

Christian Fundamentalists Teach US Children Loch Ness Monster Is Real To Disprove Evolution. The Huffington Post UK, by Lucy Sherriff. June 25, 2012.

Loch Ness Monster's Existence Being Taught at Christian Schools? By Jeff Schapiro, Christian Post Reporter, on June 26, 2012.

The Accelerated Christian "Education" program has a "science" book called *Biology 1099* which "informs" children that the Loch Ness monster is a modern-day dinosaur. Here's an excerpt:

Are dinosaurs alive today? Scientists are becoming more convinced of their existence. Have you heard of the `Loch Ness Monster' in Scotland? `Nessie,' for short has been recorded on sonar from a small submarine, described by eyewitnesses, and photographed by others. Nessie appears to be a plesiosaur.

That should sicken you. You might as well put the tooth-fairy into textbooks as proof of angels. This is stomach-churningly absurd: ***In no way should parents be allowed to decide everything that their children are taught.*** Literally the future of our planet is at stake.

CRITICAL THINKING LESSONS
FROM CHAPTER 7

Here are the takeaways from Chapter 7:

* Learn to say, "I don't know."

* It is irrational to label people based on their non-belief in non-existent beings. To do so would take an infinite amount of time to fully label each person.

* It is quicker to label people based on their positive claim.

* We should all have the right to choose—especially something so personal as a religion—without it being forced on us when we're the most vulnerable and most impressionable.

* The truth that should be taught is the one for which there is evidence.

PARTING THOUGHT:
What If You're Wrong?

Are you familiar with Pascal's Wager? Blaise Pascal was a philosopher who lived in the first half of the 17th century. He felt that there is more to be gained from assuming the existence of a supernatural being that magically created the universe than assuming the non-existence of such a being. His point seemed to be as follows:

Pascal's Wager

If there is a magical, super-powerful being that for some reason cares if you believe in its existence and who will send your non-existent "soul" to be tortured in a magical plane of existence after you die if you don't believe in Him, *then* it behooves you to believe.

Nowadays, Pascal's Wager is a question that Christians love to ask all rational non-believers, in a desperate bid to scare them into belief: *"What if you're wrong?"* the Christian asks in an ominous tone, "Do you really want to risk spending eternity in Hell?" Christians use this line of "reasoning" so much, it should be renamed *The Christian Threat of Eternal Torture.*

To respond to the wager itself:

Since your god is so fixated on torture, genocide, baby murder, mass drownings, slavery, stonings, and revenge, and is so obsessed with torturing everyone who dares to question his existence, then I must say, "No thanks" to your offer to worship your "all loving and tolerant" god. My rational mind carries a price infinitely higher than your lowball offer of fear and petty threats.

Besides, if blind fear is how we're supposed to choose our god, then there are other religions that offer *much scarier* propositions to those who don't believe.

There's an Indian religion that says if you don't believe in their supernatural universe creator, then not only will *you* suffer an eternity in their magical plane of infinite torture, but *your friends and family will suffer there also*! (This is similar to the "all loving" Christian god advocating the murder of everyone in a city where a single disbeliever lives, except this is eternally more cruel.) Not only that, but we are warned that this particular Indian Hell is *a million times more painful and agonizing than the Christian hell.* In comparison to that nasty place, the Christian hell sounds positively pleasant. Since religion is to be chosen based on avoiding misery, shouldn't all you fear-peddling Christians be advising people to believe in this Indian God instead of your own?

"No," insists the die-hard Christian, unfazed even by his own line of reasoning. "*Their* magical plane of existence where souls get tortured was just *made up* to scare people into believing. Our magical plane of existence is *real.*"

Pascal's Wager is the emptiest of threats, but ironically the question posed in the title of this essay, "What If You're Wrong?" was actually directed at believers. The risk *you guys* are taking is known as the Rational Wager. It's a bit long, so read it carefully. This is the wager that believers are taking:

Rational Wager

If there is no magical, super-powerful being and no magical plane of existence, then this one life is all you'll ever get. How many thousands of hours of your life are willing to throw away talking to your invisible sky fairy? You might as well spend that time worshiping an empty milk jug. At least the milk jug is *real*. You are quite literally praising nothing, and every moment you waste doing so is a moment you'll never get back. Let that sink in: **If there is no magical, super-powerful being then you are literally giving praise to nothing**. To rational people you seem very bizarre,

believing every word from a collection of
ancient documents written not by a magical
being but by hateful, ignorant men. Your belief
in magical trees, magical donkeys, and magical
fairies is tragic because it forces you to
abandon the awe-inspiring reality that we
actually live in. It's one thing to waste your
own life in such a sadly unthinking way, but to
subject your own children to that same kind of
empty, ignorant life is selfish and cruel. In
short, if there is no magical, super-powerful
being and no magical plane of existence, then
it behooves you to think logically, act
rationally and accept the reality of the world
you actually live in.

Don't get mad at me; I didn't write it. It's simply known as the
Rational Wager, and it's the bet every believer is taking when
he lays his money on the table of religion.

CONCLUSION

If you look back at the chapter titles in this book, you'll notice they outline what I consider the tenets of rational thought. We start by realizing that **it's healthy to question even our most fundamental beliefs**. You need to do this once in a while because there's so much at stake with every belief. By maintaining a mistaken belief, **you're missing out on something wonderful**.

The first step towards reexamining a belief is to ask yourself **what reason is there to believe that?** What unexplained phenomenon does it explain? And bear in mind, **extraordinary claims need extraordinary evidence**...or at least the same stringent evidence that all claims are held to in that arena. Regarding the incredible claims you encounter (or already hold!), be on the lookout for that commonality of complexity that all human fabrications share. Remember Occam's Razor which tells us **the simplest explanation for all the data is always the preferred choice**. Later, as you begin to rely on science to evaluate various claims, **be sure the hypothesis is falsifiable**. You can never prove a valid scientific hypothesis to be true, but the more you fail at trying to falsify it, the stronger the theory becomes. If we all made decisions and took actions based on rationally held beliefs, the world would be a whole lot better off.

* * * *

I'd like to thank each of my readers for making it all the way to the end. I hope I haven't offended, but only enlightened. Perhaps I swung too hard at times, but if I created even the tiniest crack in your armor of certainty about your beliefs, it was worth it. Life is the rarest of gifts, I hate to see people waste it.

ACKNOWLEDGMENT

I'd like to thank my wife for her infinite patience and her tireless support. This wouldn't have been possible without her. Spasibo, dorogaya maya. Ya tibya ochin silno lublu!

WORKS CITED

Anderegg, William et al. Expert Credibility in Climate Change. Proceedings of the National Academy of Sciences. April 9, 2010.

Bamford, James. "The NSA Is Building the Country's Biggest Spy Center (Watch What You Say)." *Wired*. March 15, 2012. Online.

Carter, Chris. Interview with Alex Tsakiris of Skeptico.com. *Skeptical of Skeptics, Chris Carter Tackles Near Death Experience Science*. December 7th, 2010

Cauley, Leslie. "NSA Has Massive Database of Americans' Phone Calls." *USA TODAY*. May 11th, 2006. Online.

Chung, G. S. et. al. *Obstetrician-gynecologists' Beliefs About When Pregnancy Begins*. American Journal of Obstetrics & Gynecology. Volume 206, Issue 2 , Pages 132.e1-132.e7, February 2012

CNN Wire Staff. *Medical Journal: Study Linking Autism, Vaccines is 'Elaborate Fraud'*. CNN.Com. January 6, 2011.

Cook, John. *What Evidence is There for the Hockey Stick?* SkepticalScience.Com. Nov, 2012.

- - - *Empirical Evidence That Humans Are Causing Global Warming*. SkepticalScience.Com. Nov, 2012.

Dangour, A. D., et. al. *Nutritional Quality of Organic Foods: A Systematic Review*. American Journal of Clinical Nutrition. 2009 Sep; 90(3):680-5. Epub 2009 Jul 29.

Dudley, Jonathan. *When Evangelicals Were Pro-choice*. October 30th, 2012

Dugger, Celia. "Study Cites Toll of AIDS Policy in South Africa." *New York Times*. November 25, 2008. Online.

Feuerbacher, Bjorn. and Scranton, Ryan. *Evidence for the Big Bang*. TalkOrigins.com. January 25, 2006

Garcia, Rolando. *UH Grad Student Debunks Evidence of Life in Martian Meteorite*. University of Houston, College of Natural Sciences and Mathematics. Nov, 2008. Online.

Grandia, Kevin. "The 30,000 Global Warming Petition Is Easily-Debunked Propaganda." Huffington Post. July 22, 2009. Online.

Hamilton, Jon. *A Voluble Visit with Two Talking Apes*. NPR.org. November 24, 2012

Harries, John E. et al. "Increases in Greenhouse Forcing Inferred From the Outgoing Longwave Radiation Spectra of the Earth in 1970 and 1997." *Nature*. Nature 410, 355-357. March 15, 2001.

Jeffs, William P. *New Study Adds to Finding of Ancient Life Signs in Mars Meteorite*. NASA.gov. Nov 25, 2009.

Klein, Mark. Interview with PBS Frontline, *Spying On The Homefront*. Jan 9, 2007.

LDS.org. *Book of Mormon*. Online.
<http://www.lds.org/scriptures/bofm?lang=eng>

McKay, David .S. "Origins of Magnetite Nanocrystals in Martian Meteorite ALH84001." *Journal of The Geochemical Society and The Meteoritical Society*. 18 May 2009.

Moskowitz, Clara. *Scientists Dubious Over Claim of Alien Life Evidence in Meteorite*. SPACE.com. March, 7, 2011. Online.

Muller, Robert. *A New Estimate of the Average Earth Surface Land Temperature, Spanning 1753 to 2011*. Berkeley Earth Project. Berkeleyearth.org. July, 2012.

NASA.gov. *NASA Finds 2011 Ninth-Warmest Year on Record*. NASA.gov. Jan 19, 2012.

National Academy of Sciences. "Climate Change And The Integrity of Science." *Science.* May 7, 2010.

- - - "Advancing the Science of Climate Change, 2010." NAS.edu. Division on Earth and Life Studies. Accessed online, Nov, 2012.

National Cancer Institute. *Cell Phones and Cancer Risk*. Cancer.gov. 06/18/2012

National Institute of Health. *What I need to know about Peptic Ulcers*. NIH Publication No. 11–5042. October 2010.

NOAA: *State of the Climate: 2009*. Brochure. Online. <http://www1.ncdc.noaa.gov/pub/data/cmb/bams-sotc/2009/bams-sotc-2009-brochure-lo-rez.pdf>

Offit, Paul. *Autism's False Prophets: Bad Science, Risky Medicine, and the Search For A Cure*. New York, Columbia Press. 2010.

Oreskes, Naomi. "The Scientific Consensus on Climate Change." *Science*. December 3rd, 2004: Vol. 306 no. 5702 p. 1686. Online.

Organic.org. *Organic FAQs*. November, 2012. <http://www.organic.org/education/faqs>

Organic Trade Association. *Questions and Answers About Organic*. May 15, 2012. <http://www.ota.com/organic/faq.html>

- - -. *Nutritional Considerations*. 2011.
<http://www.ota.com/organic/benefits/nutrition.html>

Park, Alice. "How Safe Are Vaccines?" *TIME*. June 02, 2008.
Online.

Prann, Elizabeth. *NSA dismisses claims Utah Data Center Watches Average Americans*. FoxNews.com. March 28, 2012.

RealityDrop.org. *Myth #85: There is No Consensus*. Popular Climate Myths. Accessed Nov. 20, 2012.

Schapiro, Jeff. *Loch Ness Monster's Existence Being Taught at Christian Schools?* Christian Post Reporter. June 26, 2012

Shermer, Michael. *Why People Believe Weird Things*. New York: Holt, 2002

Sherriff, Lucy. "Christian Fundamentalists Teach US Children Loch Ness Monster Is Real To Disprove Evolution." *The Huffington Post* UK. June 25, 2012.

Smith-Spangler, C., et. al. *Are Organic Foods Safer or Healthier Than Conventional Alternatives?: A Systematic Review*. Annals of Internal Medicine. September 4th, 2012, Vol 157, No. 5.

Taylor, Brent. et al. "Autism and Measles, Mumps, and Rubella Vaccine: No Epidemiological Evidence for a Causal Association." *The Lancet*, Volume 353, Issue 9169, Pages 2026 - 2029, 12 June 1999

Thomson, J. T. *Why We Believe in God(s): A Concise Guide to the Science of Faith*. Charlottesville: Pitchstone. 2011.

Veneziano, Gabriele. "The Myth of the Beginning of Time." *Scientific American*. Online. April 26, 2004.

Wise, Steven. *Drawing the Line*. Cambridge, MA. Perseus Books. 2002.

Zap, Claudine. *Loch Ness Monster Used to Debunk Evolution in State-funded School.* Yahoo News. June 25, 2012.
Zimmer, Carl. "Life on Mars?" *Smithsonian Magazine*, May 2005. Online.

CONTACT

To contact Mr. Milton, please send all correspondence to:
WhatIfYoureWrong@gmail.com

Please keep initial correspondences to less than 100 words.
Threats of any kind will not be tolerated, and will be forwarded
to the FBI. (And remember: An all-powerful god does not need
you defending him.)

www.ingramcontent.com/pod-product-compliance
Lightning Source LLC
Chambersburg PA
CBHW051446170526
45166CB00001B/131